From the reviews

"A remarkable book about the economic future of the United States."
—*National Review*

"By far the most trenchant analysis of a phenomenon that, if the author is correct, may be the key to our economic growth and continued prosperity."

—*New Times*

"The first book that looks at entrepreneurship as a *practice* and as such should be necessary reading for practicing executives."
—*Dallas Morning News*

"Our most enduring commentator on the practice of management and the economic institutions of society."

—*Business Week*

". . . contains a wealth of worthwhile ideas that challenge common assumptions about how businesses and organizations succeed or fail. Perhaps no one is more eminently qualified to do the job of challenging than Drucker, whose pioneering management books four decades ago have endured as classics to this day."

—*Los Angeles Times*

"Drucker believes entrepreneurship is not only possible in all institutions, it is essential to their survival. Just how to manage entrepreneurship is what this new book is all about."

—*Venture*

Books by Peter F. Drucker

MANAGEMENT

Managing the Non-Profit Organization
The Frontiers of Management
Innovation and Entrepreneurship
The Changing World of the Executive
Managing in Turbulent Times
Management: Tasks, Responsibilities, Practices
Technology, Management and Society
The Effective Executive
Managing for Results
The Practice of Management
Concept of the Corporation

ECONOMICS, POLITICS, SOCIETY

Post Capitalist Society
The New Realities
Toward the Next Economics
The Unseen Revolution
Men, Ideas and Politics
The Age of Discontinuity
Landmarks of Tomorrow
America's Next Twenty Years
The New Society
The Future of Industrial Man
The End of Economic Man

FICTION

The Temptation to Do Good
The Last of All Possible Worlds

AUTOBIOGRAPHY

Adventures of a Bystander

PETER F. DRUCKER

INNOVATION AND ENTRE-PRENEURSHIP

Practice and Principles

HARPER

NEW YORK · LONDON · TORONTO · SYDNEY

HARPER

A hardcover edition of this book was published by Harper & Row, Publishers, Inc.

HarperCollins books may be purchased for educational, business, or sales promotional use. For information, please e-mail the Special Markets Department at SPsales@harpercollins.com.

First Perennial Library edition published 1986. First Harper Business edition published 1993.

Designed by: Sidney Feinberg

Library of Congress Cataloging-in-Publication Data
Drucker, Peter Ferdinand, 1909-
   Innovation and entrepreneurship.
   "Perennial Library."
   Includes index.
   1. Small business—United States. 2. New business enterprises—United States.
3. Entrepreneur.
I. Title.
HD2346.U5D78   1986   658.4'2   84-48593
ISBN-13: 978-0-06-085113-2 (pbk.)
ISBN-10: 0-06-085113-9 (pbk.)

24 25 26 27 28 LBC 25 24 23 22 21

Contents

Preface

This book presents innovation and entrepreneurship as a practice and a discipline. It does not talk of the psychology and the character traits of entrepreneurs; it talks of their actions and behavior. It uses cases, but primarily to exemplify a point, a rule, or a warning, rather than as success stories. The work thus differs, in both intention and execution, from many of the books and articles on innovation and entrepreneurship that are being published today. It shares with them the belief in the importance of innovation and entrepreneurship. Indeed, it considers the emergence of a truly entrepreneurial economy in the United States during the last ten to fifteen years the most significant and hopeful event to have occurred in recent economic and social history. But whereas much of today's discussion treats entrepreneurship as something slightly mysterious, whether gift, talent, inspiration, or "flash of genius," this book represents innovation and entrepreneurship as purposeful tasks that can be organized—are in need of being organized—and as systematic work. It treats innovation and entrepreneurship, in fact, as part of the executive's job.

This is a practical book, but it is not a "how-to" book. Instead, it deals with the what, when, and why; with such tangibles as policies and decisions; opportunities and risks; structures and strategies; staffing, compensation, and rewards.

Innovation and entrepreneurship are discussed under three main headings: The Practice of Innovation; The Practice of Entrepreneurship; and Entrepreneurial Strategies. Each of these is an "aspect" of innovation and entrepreneurship rather than a stage.

Part I on the Practice of Innovation presents innovation alike as purposeful and as a discipline. It shows first where and how the entrepreneur searches for innovative opportunities. It then discusses the

Do's and Dont's of developing an innovative idea into a viable business or service.

Part II, The Practice of Entrepreneurship, focuses on the institution that is the carrier of innovation. It deals with entrepreneurial management in three areas: the existing business; the public-service institution; and the new venture. What are the policies and practices that enable an institution, whether business or public-service, to be a successful entrepreneur? How does one organize and staff for entrepreneurship? What are the obstacles, the impediments, the traps, the common mistakes? The section concludes with a discussion of individual entrepreneurs, their roles and their decisions.

Finally, Part III, Entrepreneurial Strategies, talks of bringing an innovation successfully to market. The test of an innovation, after all, lies not its novelty, its scientific content, or its cleverness. It lies in its success in the marketplace.

These three parts are flanked by an Introduction that relates innovation and entrepreneurship to the economy, and by a Conclusion that relates them to society.

Entrepreneurship is neither a science nor an art. It is a practice. It has a knowledge base, of course, which this book attempts to present in organized fashion. But as in all practices, medicine, for instance, or engineering, knowledge in entrepreneurship is a means to an end. Indeed, what constitutes knowledge in a practice is largely defined by the ends, that is, by the practice. Hence a book like this should be backed by long years of practice.

My work on innovation and entrepreneurship began thirty years ago, in the mid-fifties. For two years, then, a small group met under my leadership at the Graduate Business School of New York University every week for a long evening's seminar on Innovation and Entrepreneurship. The group included people who were just launching their own new ventures, most of them successfully. It included mid-career executives from a wide variety of established, mostly large organizations: two big hospitals; IBM and General Electric; one or two major banks; a brokerage house; magazine and book publishers; pharmaceuticals; a worldwide charitable organization; the Catholic Archdiocese of New York and the Presbyterian Church; and so on.

The concepts and ideas developed in this seminar were tested by its members week by week during those two years in their own work and their own institutions. Since then they have been tested, validated, refined, and revised in more than twenty years of my own consulting work. Again, a wide variety of institutions has been involved. Some were businesses, including high-tech ones such as pharmaceuticals and computer companies; "no-tech" ones such as casualty insurance companies; "world-class" banks, both American and European; one-man startup ventures; regional wholesalers of building products; and Japanese multinationals. But a host of "nonbusinesses" also were included: several major labor unions; major community organizations such as the Girl Scouts of the U.S.A. or C.A.R.E., the international relief and development cooperative; quite a few hospitals; universities and research labs; and religious organizations from a diversity of denominations.

Because this book distills years of observation, study, and practice, I was able to use actual "mini-cases," examples and illustrations both of the right and the wrong policies and practices. Wherever the name of an institution is mentioned in the text, it has either never been a client of mine (e.g., IBM) and the story is in the public domain, or the institution itself has disclosed the story. Otherwise organizations with whom I have worked remain anonymous, as has been my practice in all my management books. But the cases themselves report actual events and deal with actual enterprises.

Only in the last few years have writers on management begun to pay much attention to innovation and entrepreneurship. I have been discussing aspects of both in all my management books for decades. Yet this is the first work that attempts to present the subject in its entirety and in systematic form. This is surely a first book on a major topic rather than the last word—but I do hope it will be accepted as a seminal work.

Claremont, California
Christmas 1984

Introduction:
The Entrepreneurial Economy

I

Since the mid-seventies, such slogans as "the no-growth economy," the "deindustrialization of America," and a long-term "Kondratieff stagnation of the economy" have become popular and are invoked as if axioms. Yet the facts and figures belie every one of these slogans. What is happening in the United States is something quite different: a profound shift from a "managerial" to an "entrepreneurial" economy.

In the two decades 1965 to 1985, the number of Americans over sixteen (thereby counted as being in the work force under the conventions of American statistics) grew by two-fifths, from 129 to 180 million. But the number of Americans in paid jobs grew in the same period by one-half, from 71 to 106 million. The labor force growth was fastest in the second decade of that period, the decade from 1974 to 1984, when total jobs in the American economy grew by a full 24 million.

In no other peacetime period has the United States created as many new jobs, whether measured in percentages or in absolute numbers. And yet the ten years that began with the "oil shock" in the late fall of 1973 were years of extreme turbulence, of "energy crises," of the near-collapse of the "smokestack" industries, and of two sizable recessions.

The American development is unique. Nothing like it has happened yet in any other country. Western Europe during the period 1970 to 1984 actually *lost* jobs, 3 to 4 million of them. In 1970, western Europe still had 20 million more jobs than the United States; in 1984, it had almost 10 million less. Even Japan did far less well in job creation than the United States. During the twelve years from 1970 through 1982, jobs in Japan grew by a mere 10 percent, that is, at less than half the U.S. rate.

But America's performance in creating jobs during the seventies and early eighties also ran counter to what every expert had predicted twenty-five years ago. Then most labor force analysts expected the economy, even at its most rapid growth, to be unable to provide jobs for all the boys of the "baby boom" who were going to reach working age in the seventies and early eighties—the first large cohorts of "baby boom" babies having been born in 1949 and 1950. Actually, the American economy had to absorb twice that number. For—something nobody even dreamed of in 1970—married women began to rush into the labor force in the mid-seventies. The result is that today, in the mid-eighties, every other married woman with young children holds a paid job, whereas only one out of every five did so in 1970. And the American economy found jobs for these, too, in many cases far better jobs than women had ever held before.

And yet "everyone knows" that the seventies and early eighties were periods of "no growth," of stagnation and decline, of a "deindustrializing America," because everyone still focuses on what were the growth areas in the twenty-five years after World War II, the years that came to an end around 1970.

In those earlier years, America's economic dynamics centered in institutions that were already big and were getting bigger: the Fortune 500, that is, the country's largest businesses; governments, whether federal, state, or local; the large and super-large universities; the large consolidated high school with its six thousand or more students; and the large and growing hospital. These institutions created practically all the new jobs provided in the American economy in the quarter century after World War II. And in every recession during this period, job loss and unemployment occurred predominantly in small institutions and, of course, mainly in small businesses.

But since the late 1960s, job creation and job growth in the United States have shifted to a new sector. The old job creators have actually *lost* jobs in these last twenty years. Permanent jobs (not counting recession unemployment) in the Fortune 500 have been shrinking steadily year by year since around 1970, at first slowly, but since 1977 or 1978 at a pretty fast clip. By 1984, the Fortune 500 had lost permanently at least 4 to 6 million jobs. And governments in America, too, now employ fewer people than they did ten or fifteen years ago, if only because the number of schoolteachers has been falling as school enrollment dropped in the wake of the "baby bust" of the early sixties. Universities grew until 1980; since then, employment there has been declining. And in the early eighties, even hospital employment stopped increasing. In other words, we have not in fact created 35 million new jobs; we have created 40 million or more, since we had to offset a permanent job shrinkage of at least 5 million jobs in the traditional employing institutions. And all these new jobs must have been created by small and medium-sized institutions, most of them small and medium-sized businesses, and a great many of them, if not the majority, *new* businesses that did not even exist twenty years ago. According to *The Economist*, 600,000 new businesses are being started in the United States every year now—about seven times as many as were started in each of the boom years of the fifties and sixties.

II

"Ah," everybody will say immediately, "high tech." But things are not quite that simple. Of the 40 million-plus jobs created since 1965 in the economy, high technology did not contribute more than 5 or 6 million. High tech thus contributed no more than "smokestack" lost. All the additional jobs in the economy were generated elsewhere. And only one or two out of every hundred new businesses—a total of ten thousand a year—are remotely "high-tech," even in the loosest sense of the term.

We are indeed in the early stages of a major technological transformation, one that is far more sweeping than the most ecstatic of the "futurologists" yet realize, greater even than *Megatrends* or *Future Shock*. Three hundred years of technology came to an end after World War II. During those three centuries the model for technology was a mechanical one: the events that go on inside a star such as the sun. This period began when an otherwise almost unknown French physicist, Denis Papin,* envisaged the steam engine around 1680. They ended when we replicated in the nuclear explosion the events inside a star. For these three centuries advance in technology meant—as it does in mechanical processes—more speed, higher temperatures, higher pressures. Since the end of World War II, however, the model of technology

*The dates of all persons mentioned in the text will be found in the Index.

has become the biological process, the events inside an organism. And in an organism, processes are not organized around energy in the physicist's meaning of the term. They are organized around information.

There is no doubt that high tech, whether in the form of computers or telecommunication, robots on the factory floor or office automation, biogenetics or bioengineering, is of immeasurable qualitative importance. High tech provides the excitement and the headlines. It creates the vision for entrepreneurship and innovation in the community, and the receptivity for them. The willingness of young, highly trained people to go to work for small and unknown employers rather than for the giant bank or the worldwide electrical equipment maker is surely rooted in the mystique of "high tech"—even though the overwhelming majority of these young people work for employers whose technology is prosaic and mundane. High tech also probably stimulated the astonishing transformation of the American capital market from near-absence of venture capital as recently as the mid-sixties to near-surplus in the mid-eighties. High tech is thus what the logicians used to call the *ratio cognoscendi,* the reason why we perceive and understand a phenomenon rather than the explanation of its emergence and the cause of its existence.

Quantitatively, as has already been said, high tech is quite small still, accounting for not much more than one-eighth of the new jobs. Nor will it become much more important in terms of new jobs within the near future. Between now and the year 2000, no more than one-sixth of the jobs we can expect to create in the American economy will be high-tech jobs in all likelihood. In fact, if high tech were, as most people think, the entrepreneurial sector of the U.S. economy, then we would indeed face a "no-growth" period and a period of long-term stagnation in the trough of a "Kondratieff wave."

The Russian economist Nikolai Kondratieff was executed on Stalin's orders in the mid-1930s because his econometric model predicted, accurately as it turned out, that collectivization of Russian agriculture would lead to a sharp decline in farm production. The "fifty-year Kondratieff cycle" was based on the inherent dynamics of technology. Every fifty years, so Kondratieff asserted, a long technological wave crests. For the last twenty years of this cycle, the growth industries of the last technological advance seem to be doing exceptionally well. But what look like record profits are actually repayments of capital which is no longer needed in industries that have ceased to grow. This situation never lasts longer than twenty years, then there is a sudden crisis, usually signaled by some sort of panic. There follow twenty years of stagnation, during which the new, emerging technologies cannot generate enough jobs to make the economy itself grow again—and no one, least of all government, can do much about this.*

The industries that fueled the long economic expansion after World War II—automobiles, steel, rubber, electrical apparatus, consumer electronics, telephone, but also petroleum†—perfectly fit the Kondratieff cycle. Technologically, all of them go back to the fourth quarter of the nineteenth century or, at the very latest, to before World War I. In none of them has there been a significant breakthrough since the 1920s, whether in technology or in business concepts. When the economic growth began after World War II, they were all thoroughly mature industries. They could expand and create jobs with relatively little new capital investment, which explains why they could pay skyrocketing wages and workers' benefits and simultaneously show record profits. Yet, as Kondràtieff had predicted, these signs of robust health were as deceptive as the flush on a consumptive's cheek. The industries were corroding from within. They did not become stagnant or decline slowly. Rather, they collapsed as soon as the "oil shocks" of 1973 and 1979 dealt them the first blows. Within a few years they went from record profits to near-bankruptcy. As soon became abundantly clear, they will not be able to return to their earlier employment levels for a long time, if ever.

The high-tech industries, too, fit Kondratieff's theory. As Kondratieff had predicted, they have so far not been able to generate more jobs than the old industries have been losing. All projections indicate that they will not do much more for long years to come, at least for the rest of the century. Despite the explosive growth of computers, for instance, data processing and information handling in all their phases (design and engineering of both hardware and software, production, sales and ser-

*Kondratieff's long-wave cycle was popularized in the West by the Austro-American economist Joseph Schumpeter, in his monumental book *Business Cycles* (1939). Kondratieff's best known, most serious, and most important disciple today—and also the most serious and most knowledgeable of the prophets of "long-term stagnation"—is the MIT scientist Jay Forrester.

†Which, contrary to common belief, was the first one to start declining. In fact, petroleum ceased to be a growth industry around 1950. Since then the incremental unit of petroleum needed for an additional unit of output, whether in manufacturing, in transportation, or in heating and air conditioning, has been falling—slowly at first but rapidly since 1973.

vice) are not expected to add as many jobs to the American economy in the late 1980s and early 1990s as the steel and automotive industries are almost certain to lose.

But the Kondratieff theory fails totally to account for the 40 million jobs which the American economy actually did create. Western Europe, to be sure, has so far been following the Kondratieff script. But not the United States, and perhaps not Japan either. Something in the United States offsets the Kondratieff "long wave of technology." Something has already happened that is incompatible with the theory of long-term stagnation.

Nor does it appear at all likely that we have simply postponed the Kondratieff cycle. For in the next twenty years the need to create new jobs in the U.S. economy will be a great deal lower than it has been in the last twenty years, so that economic growth will depend far less on job creation. The number of new entrants into the American work force will be up to one-third smaller for the rest of the century—and indeed through the year 2010—than it was in the years when the children of the "baby boom" reached adulthood, that is, 1965 until 1980 or so. Since the "baby bust" of 1960–61, the birth cohorts have been 30 percent lower than they were during the "baby boom" years. And with the labor force participation of women under fifty already equal to that of men, additions to the number of women available for paid jobs will from now on be limited to natural growth, which means that they will also be down by about 30 percent.

For the future of the traditional "smokestack" industries, the Kondratieff theory must be accepted as a serious hypothesis, if not indeed as the most plausible of the available explanations. And as far as the inability of new high-tech industries to offset the stagnation of yesterday's growth industries is concerned, Kondratieff again deserves to be taken seriously. For all their tremendous qualitative importance as vision makers and pacesetters, quantitatively the high-tech industries represent tomorrow rather than today, especially as creators of jobs. They are the makers of the future rather than the makers of the present.

But as a theory of the American economy that can explain its behavior and predict its direction, Kondratieff can be considered disproven and discredited. The 40 million new jobs created in the U.S. economy during a "Kondratieff long-term stagnation" cannot be explained in Kondratieff's terms.

I do not mean to imply that there are no economic problems or dangers. Quite the contrary. A major shift in the technological foundations of the economy such as we are experiencing in the closing quarter of the twentieth century surely presents tremendous problems, economic, social, and political. We are also in the throes of a major political crisis, the crisis of that great twentieth-century success the Welfare State, with the attendant danger of an uncontrolled and seemingly uncontrollable but highly inflationary deficit. There is surely sufficient danger in the international economy, with the world's rapidly industrializing nations, such as Brazil or Mexico, suspended between rapid economic takeoff and disastrous crash, to make possible a prolonged global depression of 1930 proportions. And then there is the frightening specter of the runaway armaments race. But at least one of the fears abroad these days, that of a Kondratieff stagnation, can be considered more a figment of the imagination than reality for the United States. There we have a new, an entrepreneurial economy.

It is still too early to say whether the entrepreneurial economy will remain primarily an American phenomenon or whether it will emerge in other industrially developed countries. In Japan, there is good reason to believe that it is emerging, albeit in its own, Japanese form. But whether the same shift to an entrepreneurial economy will occur in western Europe, no one can yet say. Demographically, western Europe lags some ten to fifteen years behind America: both the "baby boom" and the "baby bust" came later in Europe than in the United States. Equally, the shift to much longer years of schooling started in western Europe some ten years later than in the United States or in Japan; and in Great Britain it has barely started yet. If, as is quite likely, demographics has been a factor in the emergence of the entrepreneurial economy in the United States, we could well see a similar development in Europe by 1990 or 1995. But this is speculation. So far, the entrepreneurial economy is purely an American phenomenon.

III

Where did all the new jobs come from? The answer is from anywhere and nowhere; in other words, from no one single source.

The magazine *Inc.*, published in Boston, has printed each year since 1982 a list of the one hundred fastest-growing, publicly owned American companies more than five years and less than fifteen years old.

Being confined to publicly owned companies, the list is heavily biased toward high tech, which has easy access to underwriters, to stock market money, and to being traded on one of the stock exchanges or over the counter. High tech is fashionable. Other new ventures, as a rule, can go public only after long years of seasoning, and of showing profits for a good deal more than five years. Yet only one-quarter of the "*Inc.* 100" are high-tech; three-quarters remain most decidedly "low-tech," year after year.

In 1982, for instance, there were five restaurant chains, two women's wear manufacturers, and twenty health-care providers on the list, but only twenty to thirty high-tech companies. And whilst America's newspapers in 1982 ran one article after the other bemoaning the "deindustrialization of America," a full half of the *Inc.* firms were manufacturing companies; only one-third were in services. Although word had it in 1982 that the Frost Belt was dying, with the Sun Belt the only possible growth area, only one-third of the "*Inc.* 100" that year were in the Sun Belt. New York had as many of these fast-growing, young, publicly owned companies as California or Texas. And Pennsylvania, New Jersey, and Massachusetts—while supposedly dying, if not already dead—also had as many as California or Texas, and as many as New York. Snowy, Minnesota, had seven. The *Inc.* lists for 1983 and 1984 showed a very similar distribution, in respect both to industry and to geography.

In 1983, the first and second companies on another *Inc.* list—the "*Inc.* 500" list of fast-growing, young, privately held companies—were, respectively, a building contractor in the Pacific Northwest (in a year in which construction was supposedly at an all-time low) and a California manufacturer of physical exercise equipment for the home.

Any inquiry among venture capitalists yields the same pattern. Indeed, in their portfolios, high tech is usually even less prominent. The portfolio of one of the most successful venture capital investors does include several high-tech companies: a new computer software producer, a new venture in medical technology, and so on. But the most profitable investment in this portfolio, the new company that has been growing the fastest in both revenues and profitability during the three years 1981–83, is that most mundane and least high-tech of businesses, a chain of barbershops. And next to it, both in sales growth and profitability, comes a chain of dentistry offices, followed by a manufacturer of handtools and by a finance company that leases machinery to small businesses.

Among the businesses I know personally, the one that has created the most jobs during the five years 1979–84, and has also grown the fastest in revenues and profits, is a financial services firm. Within five years this firm alone has created two thousand new jobs, most of them exceedingly well paid. Though a member of the New York Stock Exchange, only about one-eighth of its business is in stocks. The rest is in annuities, tax-exempt bonds, money-market funds and mutual funds, mortgage-trust certificates, tax-shelter partnerships, and a host of similar investments for what the firm calls "the intelligent investor." Such investors are defined as the well-to-do but not rich professional, small businessman, or farmer, in small towns or in the suburbs, who makes more money than he spends and thus looks for places to put his savings, but who is also realistic enough not to expect to become rich through investment.

The most revealing source of information about the growth sectors of the U.S. economy I have been able to find is a study of the one hundred fastest-growing "mid-size" companies, that is, companies with revenues of between $25 million and $1 billion. This study was conducted during 1981–83 for the American Business Conference by two senior partners of McKinsey & Company, the consulting firm.*

These mid-sized growth companies grew at three times the rate of the Fortune 500 in sales and in profits. The Fortune 500 have been losing jobs steadily since 1970. But these mid-sized growth companies added jobs between 1970 and 1983 at three times the rate of job growth in the entire U.S. economy. Even in the depression years 1981–82 when jobs in U.S. industry declined by almost 2 percent, the hundred mid-sized growth companies increased their employment by one full percentage point. The companies span the economic spectrum. There are high-tech ones among them, to be sure. But there are also financial services companies—the New York investment and brokerage firm of Donaldson, Lufkin & Jenrette, for instance. One of the best performers in the group is a company making and selling living-room furniture; another one is making and marketing doughnuts; a third, high-quality chinaware; a fourth, writing instruments; a fifth, household paints; a

---

*It was published under the title "Lessons from America's Mid-sized Growth Companies," by Richard E. Cavenaugh and Donald K. Clifford, Jr., in the Autumn 1983 issue of the *McKinsey Quarterly*.

sixth has expanded from printing and publishing local newspapers into consumer marketing services; a seventh produces yarns for the textile industry; and so forth. And where "everybody knows" that growth in the American economy is exclusively in services, more than half of these "mid-sized growth" companies are in manufacturing.

To make things more confusing still, the growth sector of the U.S. economy during the last ten to fifteen years, while entirely nongovernmental, includes a fairly large and growing number of enterprises that are not normally considered businesses, though quite a few are now being organized as profit-making companies. The most visible of these are, of course, in the health-care field. The traditional American community hospital is in deep trouble these days. But there are fast-growing and flourishing hospital chains, both "profit" and (increasingly) "not-for-profit" ones. Even faster growing are the "freestanding" health facilities, such as hospices for the terminally ill, medical and diagnostic laboratories, freestanding surgery centers, freestanding maternity homes, psychiatric "walk-in" clinics, or centers for geriatric diagnosis and treatment.

The public schools are shrinking in almost every American community. But despite the decline in the total number of children of school age as a result of the "baby bust" of the 1960s, a whole new species of non-profit but private schools is flourishing. In the small California city in which I live, a neighborhood babysitting cooperative, founded around 1980 by a few mothers for their own children, had by 1984 grown into a school with two hundred students going on into the fourth grade. And a "Christian" school founded a few years ago by the local Baptists is taking over from the city of Claremont a junior high school built fifteen years ago and left standing vacant for lack of pupils for the last five years. Continuing education of all kinds, whether in the form of executive management programs for mid-career managers or refresher courses for doctors, engineers, lawyers, and physical therapists, is booming; even during the severe 1982–83 recession, such programs suffered only a short setback.

One additional area of entrepreneurship, and a very important one, is the emerging "Fourth Sector" of public-private partnerships in which government units, either states or municipalities, determine performance standards and provide the money. But then they contract out a service—fire protection, garbage collection, or bus transportation—to a private business on the basis of competitive bids, thus ensuring both better service and substantially lower costs. The city of Lincoln, Nebraska, has been a pioneer in this area since Helen Boosalis was first elected mayor in 1975—the same Lincoln, Nebraska, where a hundred years ago the Populists and William Jennings Bryan first started us on the road to municipal ownership of public services. Pioneering work in this area is also being done in Texas—in San Antonio and in Houston, for instance—and especially in Minneapolis at the Hubert Humphrey Institute of the University of Minnesota. Control Data Corporation, a leading computer manufacturer also in Minneapolis, is building public-private partnerships in education and even in the management and rehabilitation of prisoners. And if there is one action that can save the postal service in the long run—for surely there is a limit to the public's willingness to pay ever larger subsidies and ever higher rates for ever-shrinking service—it may be the contracting out of first-class service (or what's still left of it ten years hence) to the "Fourth Sector," through competitive bids.

I V

Is there anything at all that these growth enterprises have in common other than growth and defiance of the Kondratieff stagnation? Actually, they are all examples of "new technology," all new applications of knowledge to human work, which is, after all, the definition of technology. Only the "technology" is not electronics or genetics or new materials. The "new technology" is entrepreneurial management.

Once this is seen, then the astonishing job growth of the American economy during the last twenty, and especially the last ten years can be explained. It can even be reconciled with the Kondratieff theory. The United States—and to some extent also Japan—is experiencing what might be called an "atypical Kondratieff cycle."

Since Joseph Schumpeter first pointed it out in 1939, we have known that what actually happened in the United States and in Germany in the fifty years between 1873 and World War I does not fit the Kondratieff cycle. The first Kondratieff cycle, based on the railway boom, came to an end with the crash of the Vienna Stock Exchange in 1873, a crash that brought down stock exchanges worldwide and ushered in a severe depression. Great Britain and France did then enter a long period of industrial stagnation during which the new emerging technologies—steel, chemicals, electrical apparatus, telephone, and finally, automobiles—could not create enough jobs to offset the stagnation in the old industries, such as railway construction, coal mining, or textiles.

But this did not happen in the United States or in Germany, nor indeed in Austria, despite the traumatic impact of the Viennese stock market crash from which Austrian politics never quite recovered. These countries were severely jolted at first. Five years later they had pulled out of the slump and were growing again, fast. In terms of "technology," these countries were no different from stagnating Britain or France. What explains their different economic behavior was one factor, and one factor only: the entrepreneur. In Germany, for instance, the single most important economic event in the years between 1870 and 1914 was surely the creation of the Universal Bank. The first of these, the Deutsche Bank, was founded by Georg Siemens in 1870* with the specific mission of finding entrepreneurs, financing entrepreneurs, and forcing upon them organized, disciplined management. In the economic history of the United States the entrepreneurial bankers such as J. P. Morgan in New York played a similar role.

Today, something very similar seems to be happening in the United States and perhaps also to some extent in Japan.

Indeed, high tech is the one sector that is not part of this new "technology," this "entrepreneurial management." The Silicon Valley high-tech entrepreneurs still operate mainly in the nineteenth-century mold. They still believe in Benjamin Franklin's dictum: "If you invent a better mousetrap the world will beat a path to your door." It does not yet occur to them to ask what makes a mousetrap "better" or for whom?

There are, of course, plenty of exceptions, high-tech companies that know well how to manage entrepreneurship and innovation. But then there were exceptions during the nineteenth century, too. There was the German, Werner Siemens, who founded and built the company that still bears his name. There was George Westinghouse, the American, a great inventor but also a great business builder, who left behind two companies that still bear his name, one a leader in the field of transportation, the other a major force in the electrical apparatus industry.

But for the "high-tech" entrepreneur, the archetype still seems to be Thomas Edison. Edison, the nineteenth century's most successful inventor, converted invention into the discipline we now call research. His real ambition, however, was to be a business builder and to become

---

*On Georg Siemens and the Universal Bank, see Chapter 9.

a tycoon. Yet he so totally mismanaged the businesses he started that he had to be removed from every one of them to save it. Much, if not most high tech is still being managed, or more accurately mismanaged, Edison's way.

This explains, first, why the high-tech industries follow the traditional pattern of great excitement, rapid expansion, and then sudden shakeout and collapse, the pattern of "from rags to riches and back to rags again" in five years. Most of Silicon Valley—but most of the new biological high-tech companies as well—are still inventors rather than innovators, still speculators rather than entrepreneurs. And this, too, perhaps explains why high tech so far conforms to the Kondratieff prediction and does not generate enough jobs to make the whole economy grow again.

But the "low tech" of systematic, purposeful, managed entrepreneurship does.

V

Of all the major modern economists only Joseph Schumpeter concerned himself with the entrepreneur and his impact on the economy. Every economist knows that the entrepreneur is important and has impact. But, for economists, entrepreneurship is a "meta-economic" event, something that profoundly influences and indeed shapes the economy without itself being part of it. And so too, for economists, is technology. Economists do not, in other words, have any explanation as to why entrepreneurship emerged as it did in the late nineteenth century and as it seems to be doing again today, nor why it is limited to one country or to one culture. Indeed, the events that explain why entrepreneurship becomes effective are probably not in themselves economic events. The causes are likely to lie in changes in values, perception, and attitude, changes perhaps in demographics, in institutions (such as the creation of entrepreneurial banks in Germany and the United States around 1870), perhaps changes in education as well.

Something, surely, has happened to young Americans—and to fairly large numbers of them—to their attitudes, their values, their ambitions, in the last twenty to twenty-five years. Only it is clearly not what anyone looking at the young Americans of the late 1960s could possibly have predicted. How do we explain, for instance, that all of a sudden there are such large numbers of people willing both to work like demons for long years and to choose grave risks rather than big organization security? Where are the hedonists, the status seekers, the "me-too-ers," the conformists? Conversely, where are all the young people who, we were told fifteen years ago, were turning their backs on material values, on money, goods, and worldly success, and were going to restore to America a "laid-back," if not a pastoral "greenness"? Whatever the explanation, it does not fit in with what all the soothsayers of the last thirty years—David Riesman in *The Lonely Crowd*, William H. Whyte in *The Organization Man*, Charles Reich in *The Greening of America*, or Herbert Marcuse—predicted about the younger generation. Surely the emergence of the entrepreneurial economy is as much a cultural and psychological as it is an economic or technological event. Yet whatever the causes, the effects are above all economic ones.

And the vehicle of this profound change in attitudes, values, and above all in behavior is a "technology." It is called management. What has made possible the emergence of the entrepreneurial economy in America is new applications of management:

— to new enterprises, whether businesses or not, whereas most people until now have considered management applicable to existing enterprises only;

— to small enterprises, whereas most people were absolutely sure only a few years ago that management was for the "big boys" only;

— to nonbusinesses (health care, education, and so on), whereas most people still hear "business" when they encounter the word "management";

— to activities that were simply not considered to be "enterprises" at all, such as local restaurants;

— and above all, to systematic innovation: to the search for and the exploitation of new opportunities for satisfying human wants and human needs.

As a "useful knowledge," a *techné*, management is the same age as the other major areas of knowledge that underlie today's high-tech industries, whether electronics, solid-state physics, genetics, or immunology. Management's roots lie in the time around World War I. Its early shoots came up in the mid-1920s. But management is a "useful knowledge" like engineering or medicine, and as such it first had to develop as a practice before it could become a discipline. By the late

1930s, there were a few major enterprises around—at that time mostly businesses—that practiced "management" in the United States: the DuPont Company and its half brother, General Motors, but also a large retailer, Sears, Roebuck. On the other side of the Atlantic there was Siemens in Germany, or the department store chain of Marks and Spencer in Great Britain. But management as a discipline originated during and right after World War II.*

Beginning around 1955, the entire developed world experienced a "management boom."† The social technology we call management was first presented to the general public, including managers themselves, some forty years ago. It then rapidly became a discipline rather than the hit-or-miss practice of a few isolated true believers. And in these forty years management has had as much impact as any of the "scientific breakthroughs" of the period—perhaps a good deal more. It may not be solely or even primarily responsible for the fact that society in every single developed country has become since World War II a society of organizations. It may not be solely or even primarily responsible for the fact that in every developed society today the great majority of people —and the overwhelming majority of educated people—work as employees in organizations, including of course the bosses themselves, who increasingly tend to be "professional managers," that is, hired hands, rather than owners. But surely if management had not emerged as a systematic discipline, we could not have organized what is now a social reality in every developed country: the society of organizations and the "employee society."

We still have quite a bit to learn about management, admittedly, and above all about the management of the knowledge worker. But the fundamentals are reasonably well known by now. Indeed, what was an esoteric cult only forty years ago, when most executives even in large companies did not in fact realize that they practiced management, now has become commonplace.

But by and large management until recently was seen as being

---

*My first two management books, *Concept of the Corporation* (1946; a study of General Motors), and *The Practice of Management* (1954) were indeed the original attempts to organize and present management as a systematic body of knowledge, that is, as a discipline.

†This by now has even reached Communist China. One of the first actions of the Chinese government after the fall of the "Gang of Four" was to establish an Enterprise Management Agency directly responsible to the prime minister, and to import a Graduate Business School from the United States.

confined to business, and within business, to "big business." In the early seventies, when the American Management Association invited the heads of small business to its "Presidents' Course" in Management, it was told again and again: "Management? That's not for me—that's only for big companies." Up to 1970 or 1975, American hospital administrators still rejected anything that was labeled "management." "We're hospital people, not business people," they said. (In the universities the faculties are still saying the same thing even though they will simultaneously complain how "badly managed" their institution is.) And indeed for a long time, from the end of World War II until 1970, "progress" meant building bigger institutions.

This twenty-five-year trend toward building bigger organizations in every social sphere—business, labor union, hospital, school, university, and so on—had many causes. But the belief that we knew how to manage bigness and did not really know how to manage small enterprises was surely a major factor. It had, for instance, a great deal to do with the rush toward the very large consolidated American high school. "Education," it was argued, "requires professional administration, and this in turn works only in large rather than small enterprises."

During the last ten or fifteen years we have reversed this trend. In fact, we might now have a trend toward "deinstitutionalizing" America rather than one toward "deindustrializing" it. For almost fifty years, ever since the 1930s, it was widely believed in the United States and in western Europe too that the hospital was the best place for anyone not quite well, let alone for anyone seriously sick. "The sooner the patient gets to the hospital, the better care we can take of him," was the prevailing belief, shared by doctors and patients alike. In the last few years, we have been reversing this trend. We now increasingly believe that the longer we can keep patients away from the hospital and the sooner we can get them out, the better. Surely this reversal has little to do with either health care or with management. It is a reaction—whether permanent or short-lived—against the worship of centralization, of "planning," of government which began in the 1920s and 1930s, and which in the United States reached its peak in the Kennedy and Johnson administrations of the 1960s. However, we could not indulge in this "deinstitutionalization" in the health-care field if we had not acquired the competence and the confidence to manage small institutions and "non-businesses," that is, health-care institutions.

All told we are learning that management may well both be more needed and have greater impact on the small entrepreneurial organization than it has in the big "managed" one. Above all, management, we are learning now, has as much to contribute to the new, the entrepreneurial enterprise, as to the existing, ongoing "managerial" one.

To take a specific example, hamburger stands have been around in the United States since the nineteenth century; after World War II they sprang up on big-city street corners. But in the McDonald's hamburger chain—one of the success stories of the last twenty-five years—management was being applied to what had always been a hit-and-miss, mom-and-pop operation. McDonald's first designed the end product; then it redesigned the entire process of making it; then it redesigned or in many cases invented the tools so that every piece of meat, every slice of onion, every bun, every piece of fried potato would be identical, turned out in a precisely timed and fully automated process. Finally, McDonald's studied what "value" meant to the customer, defined it as quality and predictability of product, speed of service, absolute cleanliness, and friendliness, then set standards for all of these, trained for them, and geared compensation to them.

All of which is management, and fairly advanced management at that.

Management is the new technology (rather than any specific new science or invention) that is making the American economy into an entrepreneurial economy. It is also about to make America into an entrepreneurial *society*. Indeed, there may be greater scope in the United States—and in developed societies generally—for social innovation in education, health care, government, and politics than there is in business and the economy. And again, entrepreneurship in society—and it is badly needed—requires above all application of the basic concepts, the basic *techné*, of management to new problems and new opportunities.

This means that the time has now come to do for entrepreneurship and innovation what we first did for management in general some thirty years ago: to develop the principles, the practice, and the discipline.

I

THE PRACTICE
OF INNOVATION

Innovation is the specific tool of entrepreneurs, the means by which they exploit change as an opportunity for a different business or a different service. It is capable of being presented as a discipline, capable of being learned, capable of being practiced. Entrepreneurs need to search purposefully for the sources of innovation, the changes and their symptoms that indicate opportunities for successful innovation. And they need to know and to apply the principles of successful innovation.

1

Systematic Entrepreneurship

I

"The entrepreneur," said the French economist J. B. Say around 1800, "shifts economic resources out of an area of lower and into an area of higher productivity and greater yield." But Say's definition does not tell us who this "entrepreneur" is. And since Say coined the term almost two hundred years ago, there has been total confusion over the definitions of "entrepreneur" and "entrepreneurship."

In the United States, for instance, the entrepreneur is often defined as one who starts *his own, new* and *small business.* Indeed, the courses in "Entrepreneurship" that have become popular of late in American business schools are the linear descendants of the course in starting one's own small business that was offered thirty years ago, and in many cases, not very different.

But not every new small business is entrepreneurial or represents entrepreneurship.

The husband and wife who open another delicatessen store or another Mexican restaurant in the American suburb surely take a risk. But are they entrepreneurs? All they do is what has been done many times before. They gamble on the increasing popularity of eating out in their area, but create neither a new satisfaction nor new consumer demand. Seen under this perspective they are surely not entrepreneurs even though theirs is a new venture.

McDonald's, however, was entrepreneurship. It did not invent anything, to be sure. Its final product was what any decent American restaurant had produced years ago. But by applying management concepts and management techniques (asking, What is "value" to the customer?), standardizing the "product," designing process and tools, and by basing training on the analysis of the work to be done and then

21

setting the standards it required, McDonald's both drastically upgraded the yield from resources, and created a new market and a new customer. This is entrepreneurship.

Equally entrepreneurial is the growing foundry started by a husband and wife team a few years ago in America's Midwest, to heat-treat ferrous castings to high-performance specifications—for example, the axles for the huge bulldozers used to clear the land and dig the ditches for a natural gas pipeline across Alaska. The science needed is well known; indeed, the company does little that has not been done before. But in the first place the founders systematized the technical information: they can now punch the performance specifications into their computer and get an immediate printout of the treatment required. Secondly, the founders systematized the process. Few orders run to more than half a dozen pieces of the same dimension, the same metallic composition, the same weight, and the same performance specifications. Yet the castings are being produced in what is, in effect, a flow process rather than in batches, with computer-controlled machines and ovens adjusting themselves.

Precision castings of this kind used to have a rejection rate of 30 to 40 percent; in this new foundry, 90 percent or more are flawless when they come off the line. And the costs are less than two-thirds of those of the cheapest competitor (a Korean shipyard), even though the Midwestern foundry pays full American union wages and benefits. What is "entrepreneurial" in this business is not that it is new and still small (though growing rapidly). It is the realization that castings of this kind are distinct and separate; that demand for them has grown so big as to create a "market niche"; and that technology, especially computer technology, now makes possible the conversion of an art into a scientific process.

Admittedly, all new small businesses have many factors in common. But to be entrepreneurial, an enterprise has to have special characteristics over and above being new and small. Indeed, entrepreneurs are a minority among new businesses. They create something new, something different; they change or transmute values.

An enterprise also does not need to be small and new to be an entrepreneur. Indeed, entrepreneurship is being practiced by large and often old enterprises. The General Electric Company (G.E.), one of the world's biggest businesses and more than a hundred years old, has a long history of starting new entrepreneurial businesses from scratch and raising them into sizable industries. And G.E. has not confined itself to entrepreneurship in manufacturing. Its financing arm, G.E. Credit Corporation, in large measure triggered the upheaval that is transforming the American financial system and is now spreading rapidly to Great Britain and western Europe as well. G.E. Credit in the sixties ran around the Maginot Line of the financial world when it discovered that commercial paper could be used to finance industry. This broke the banks' traditional monopoly on commercial loans.

Marks and Spencer, the very large British retailer, has probably been more entrepreneurial and innovative than any other company in western Europe these last fifty years, and may have had greater impact on the British economy and even on British society, than any other change agent in Britain, and arguably more than government or laws.

Again, G.E. and Marks and Spencer have many things in common with large and established businesses that are totally unentrepreneurial. What makes them "entrepreneurial" are specific characteristics other than size or growth.

Finally, entrepreneurship is by no means confined solely to economic institutions.

No better text for a *History of Entrepreneurship* could be found than the creation and development of the modern university, and especially the modern American university. The modern university as we know it started out as the invention of a German diplomat and civil servant, Wilhelm von Humboldt, who in 1809 conceived and founded the University of Berlin with two clear objectives: to take intellectual and scientific leadership away from the French and give it to the Germans; and to capture the energies released by the French Revolution and turn them against the French themselves, especially Napoleon. Sixty years later, around 1870, when the German university itself had peaked, Humboldt's idea of the university as a change agent was picked up across the Atlantic, in the United States. There, by the end of the Civil War, the old "colleges" of the colonial period were dying of senility. In 1870, the United States had no more than half the college students it had had in 1830, even though the population had nearly tripled. But in the next thirty years a galaxy of American university presidents* created and built a new "American university"—both distinctly new

---

*See the section on The American University in my book *Management: Tasks, Responsibilities, Practices* (New York: Harper & Row, 1973), pages 150–152.

and distinctly American—which then, after World War I, soon gained
for the United States worldwide leadership in scholarship and research,
just as Humboldt's university had gained worldwide leadership in scholarship and research for Germany a century earlier.

After World War II a new generation of American academic enter-
preneurs innovated once again, building new "private" and "metropolitan" universities: Pace University, Fairleigh-Dickinson, and the New
York Institute of Technology in the New York area; Northeastern in
Boston; Santa Clara and Golden Gate on the West Coast; and so on.
They have constituted a major growth sector in American higher education in the last thirty years. Most of these new schools seem to differ
little from the older institutions in their curriculum. But they were
deliberately designed for a new and different "market"—for people in
mid-career rather than for youngsters fresh out of high school; for
big-city students commuting to the university at all hours of the day and
night rather than for students living on campus and going to school full
time, five days a week from nine to five; and for students of widely
diversified, indeed, heterogenous backgrounds rather than for the "college kid" of the American tradition. They were a response to a major
shift in the market, a shift in the status of the college degree from
"upper-class" to "middle-class," and to a major shift in what "going to
college" means. They represent entrepreneurship.

One could equally write a casebook on entrepreneurship based on
the history of the hospital, from the first appearance of the modern
hospital in the late eighteenth century in Edinburgh and Vienna, to the
creation of the various forms of the "community hospital" in nineteenth-century America, to the great specialized centers of the early
twentieth century, the Mayo Clinic or the Menninger Foundation, to
the emergence of the hospital as health-care center in the post–World
War II period. And now new entrepreneurs are busily changing the
hospital again into specialized "treatment centers": ambulatory surgical clinics, freestanding maternity centers or psychiatric centers where
the emphasis is not, as in the traditional hospital, on caring for the
patient but on specialized "needs."

Again, not every nonbusiness service institution is entrepreneurial;
far from it. And the minority that is still has all the characteristics, all
the problems, all the identifying marks of the service institution.* What

*On this, see the section Performance in the Service Institution (Chapters 11–14) in

makes these service institutions entrepreneurial is something different, something specific.

Whereas English speakers identify entrepreneurship with the new, small business, the Germans identify it with power and property, which is even more misleading. The *Unternehmer*—the literal translation into German of Say's *entrepreneur*—is the person who both owns and runs a business (the English term would be "owner-manager"). And the word is used primarily to distinguish the "boss," who also owns the business, from the "professional manager" and from "hired hands" altogether.

But the first attempts to create systematic entrepreneurship—the entrepreneurial bank founded in France in 1857 by the Brothers Pereire in their Crédit Mobilier, then perfected in 1870 across the Rhine by Georg Siemens in his Deutsche Bank, and brought across the Atlantic to New York at about the same time by the young J. P. Morgan—did not aim at ownership. The task of the banker as entrepreneur was to mobilize *other people's money* for allocation to areas of higher productivity and greater yield. The earlier bankers, the Rothschilds, for example, became owners. Whenever they built a railroad, they financed it with their own money. The entrepreneurial banker, by contrast, never wanted to be an owner. He made his money by selling to the general public the shares of the enterprises he had financed in their infancy. And he got the money for his ventures by borrowing from the general public.

Nor are entrepreneurs capitalists, although of course they need capital as do all economic (and most noneconomic) activities. They are not investors, either. They take risks, of course, but so does anyone engaged in any kind of economic activity. The essence of economic activity is the commitment of present resources to future expectations, and that means to uncertainty and risk. The entrepreneur is also not an employer, but can be, and often is, an employee—or someone who works alone and entirely by himself or herself.

Entrepreneurship is thus a distinct feature whether of an individual or of an institution. It is not a personality trait; in thirty years I have seen people of the most diverse personalities and temperaments perform

---

Management: Tasks, Responsibilities, Practices, but also Chapter 14 of this book, Entrepreneurship in the Service Institution.

well in entrepreneurial challenges. To be sure, people who need certainty are unlikely to make good entrepreneurs. But such people are unlikely to do well in a host of other activities as well—in politics, for instance, or in command positions in a military service, or as the captain of an ocean liner. In all such pursuits decisions have to be made, and the essence of any decision is uncertainty.

But everyone who can face up to decision making can learn to be an entrepreneur and to behave entrepreneurially. Entrepreneurship, then, is behavior rather than personality trait. And its foundation lies in concept and theory rather than in intuition.

II

Every practice rests on theory, even if the practitioners themselves are unaware of it. Entrepreneurship rests on a theory of economy and society. The theory sees change as normal and indeed as healthy. And it sees the major task in society—and especially in the economy—as doing something different rather than doing better what is already being done. This is basically what Say, two hundred years ago, meant when he coined the term *entrepreneur*. It was intended as a manifesto and as a declaration of dissent: the entrepreneur upsets and disorganizes. As Joseph Schumpeter formulated it, his task is "creative destruction."

Say was an admirer of Adam Smith. He translated Smith's *Wealth of Nations* (1776) into French and tirelessly propagated throughout his life Smith's ideas and policies. But his own contribution to economic thought, the concept of the entrepreneur and of entrepreneurship, is independent of classical economics and indeed incompatible with it. Classical economics optimizes what already exists, as does mainstream economic theory to this day, including the Keynesians, the Friedmanites, and the Supply-siders. It focuses on getting the most out of existing resources and aims at establishing equilibrium. It cannot handle the entrepreneur but consigns him to the shadowy realm of "external forces," together with climate and weather, government and politics, pestilence and war, but also technology. The traditional economist, regardless of school or "ism," does not deny, of course, that these external forces exist or that they matter. But they are not part of his world, not accounted for in his model, his equations, or his predictions. And while Karl Marx had the keenest appreciation of technology—he was the first and is still one of the best historians of technology—he could not admit the entrepreneur and entrepreneurship into either his system or his economics. *All* economic change in Marx beyond the optimization of present resources, that is, the establishment of equilibrium, is the result of changes in property and power relationships, and hence "politics," which places it outside the economic system itself.

Joseph Schumpeter was the first major economist to go back to Say. In his classic *Die Theorie der Wirtschaftlichen Entwicklung (The Theory of Economic Dynamics),* published in 1911, Schumpeter broke with traditional economics—far more radically than John Maynard Keynes was to do twenty years later. He postulated that dynamic disequilibrium brought by the innovating entrepreneur, rather than equilibrium and optimization, is the "norm" of a healthy economy and the central reality for economic theory and economic practice.

Say was primarily concerned with the economic sphere. But his definition only calls for the resources to be "economic." The purpose to which these resources are dedicated need not be what is traditionally thought of as economic. Education is not normally considered "economic"; and certainly economic criteria are hardly appropriate to determine the "yield" of education (though no one knows what other criteria might pertain). But the resources of education are, of course, economic. They are in fact identical with those used for the most unambiguously economic purpose such as making soap for sale. Indeed, the resources for all *social* activities of human beings are the same and are "economic" resources: capital (that is, the resources withheld from current consumption and allocated instead to future expectations), physical resources, whether land, seed corn, copper, the classroom, or the hospital bed; labor, management, and time. Hence entrepreneurship is by no means limited to the economic sphere although the term originated there. It pertains to all activities of human beings other than those one might term "existential" rather than "social." And we now know that there is little difference between entrepreneurship whatever the sphere. The entrepreneur in education and the entrepreneur in health care—both have been fertile fields—do very much the same things, use very much the same tools, and encounter very much the same problems as the entrepreneur in a business or a labor union.

Entrepreneurs see change as the norm and as healthy. Usually, they do not bring about the change themselves. But—and this defines entrepreneur and entrepreneurship—*the entrepreneur always searches for change, responds to it, and exploits it as an opportunity.*

III

Entrepreneurship, it is commonly believed, is enormously risky. And, indeed, in such highly visible areas of innovation as high tech—microcomputers, for instance, or biogenetics—the casualty rate is high and the chances of success or even of survival seem to be quite low.

But why should this be so? Entrepreneurs, by definition, shift resources from areas of low productivity and yield to areas of higher productivity and yield. Of course, there is a risk they may not succeed. But if they are even moderately successful, the returns should be more than adequate to offset whatever risk there might be. One should thus expect entrepreneurship to be considerably less risky than optimization. Indeed, nothing could be as risky as optimizing resources in areas where the proper and profitable course is innovation, that is, where the opportunities for innovation already exist. Theoretically, entrepreneurship should be the least risky rather than the most risky course.

In fact, there are plenty of entrepreneurial organizations around whose batting average is so high as to give the lie to the all but universal belief in the high risk of entrepreneurship and innovation.

In the United States, for instance, there is Bell Lab, the innovative arm of the Bell Telephone System. For more than seventy years—from the design of the first automatic switchboard around 1911 until the design of the optical fiber cable around 1980, including the invention of transistor and semiconductor, but also basic theoretical and engineering work on the computer—Bell Lab produced one winner after another. The Bell Lab record would indicate that even in the high-tech field, entrepreneurship and innovation can be low-risk.

IBM, in a fast-moving high-tech field, that of the computer, and in competition with the "old pros" in electricity and electronics, has so far not had one major failure. Nor, in a far more prosaic industry, has the most entrepreneurial of the world's major retailers, the British department store chain Marks and Spencer. The world's largest producer of branded and packaged consumer goods, Procter & Gamble, similarly has had a near-perfect record of successful innovations. And a "middle-tech" company, 3M in St. Paul, Minnesota, which has created around one hundred new businesses or new major product lines in the last sixty years, has been successful four out of every five times in its ventures. This is only a small sample of the entrepreneurs who somehow innovate at low risk. Surely there are far too many of them for low-risk entrepreneurship to be a fluke, a special dispensation of the gods, an accident, or mere chance.

There are also enough individual entrepreneurs around whose batting average in starting new ventures is so high as to disprove the popular belief of the high risk of entrepreneurship.

Entrepreneurship is "risky" mainly because so few of the so-called entrepreneurs know what they are doing. They lack the methodology. They violate elementary and well-known rules. This is particularly true of high-tech entrepreneurs. To be sure (as will be discussed in Chapter 9), high-tech entrepreneurship and innovation are intrinsically more difficult and more risky than innovation based on economics and market structure, on demographics, or even on something as seemingly nebulous and intangible as *Weltanschauung*—perceptions and moods. But even high-tech entrepreneurship need not be "high-risk," as Bell Lab and IBM prove. It does need, however, to be systematic. It needs to be managed. Above all, it needs to be based on *purposeful innovation*.

2

Purposeful Innovation and the Seven Sources for Innovative Opportunity

Entrepreneurs innovate. Innovation is the specific instrument of entrepreneurship. It is the act that endows resources with a new capacity to create wealth. Innovation, indeed, creates a resource. There is no such thing as a "resource" until man finds a use for something in nature and thus endows it with economic value. Until then, every plant is a weed and every mineral just another rock. Not much more than a century ago, neither mineral oil seeping out of the ground nor bauxite, the ore of aluminum, were resources. They were nuisances; both render the soil infertile. The penicillin mold was a pest, not a resource. Bacteriologists went to great lengths to protect their bacterial cultures against contamination by it. Then in the 1920s, a London doctor, Alexander Fleming, realized that this "pest" was exactly the bacterial killer bacteriologists had been looking for—and the penicillin mold became a valuable resource.

The same holds just as true in the social and economic spheres. There is no greater resource in an economy than "purchasing power." But purchasing power is the creation of the innovating entrepreneur.

The American farmer had virtually no purchasing power in the early nineteenth century; he therefore could not buy farm machinery. There were dozens of harvesting machines on the market, but however much he might have wanted them, the farmer could not pay for them. Then one of the many harvesting-machine inventors, Cyrus McCormick, invented installment buying. This enabled the farmer to pay for a harvesting machine out of his future earnings rather than out of past savings—and suddenly the farmer had "purchasing power" to buy farm equipment.

Equally, whatever changes the wealth-producing potential of already existing resources constitutes innovation.

There was not much new technology involved in the idea of moving a truck body off its wheels and onto a cargo vessel. This "innovation," the container, did not grow out of technology at all but out of a new perception of the "cargo vessel" as a materials-handling device rather than a "ship," which meant that what really mattered was to make the time in port as short as possible. But this humdrum innovation roughly quadrupled the productivity of the ocean-going freighter and probably saved shipping. Without it, the tremendous expansion of world trade in the last forty years—the fastest growth in any major economic activity ever recorded—could not possibly have taken place.

What really made universal schooling possible—more so than the popular commitment to the value of education, the systematic training of teachers in schools of education, or pedagogic theory—was that lowly innovation, the textbook. (The textbook was probably the invention of the great Czech educational reformer Johann Amos Comenius, who designed and used the first Latin primers in the mid-seventeenth century.) Without the textbook, even a very good teacher cannot teach more than one or two children at a time; with it, even a pretty poor teacher can get a little learning into the heads of thirty or thirty-five students.

Innovation, as these examples show, does not have to be technical, does not indeed have to be a "thing" altogether. Few technical innovations can compete in terms of impact with such social innovations as the newspaper or insurance. Installment buying literally transforms economies. Wherever introduced, it changes the economy from supply-driven to demand-driven, regardless almost of the productive level of the economy (which explains why installment buying is the first practice that any Marxist government coming to power immediately suppresses: as the Communists did in Czechoslovakia in 1948, and again in Cuba in 1959). The hospital, in its modern form a social innovation of the Enlightenment of the eighteenth century, has had greater impact on health care than many advances in medicine. Management, that is, the "useful knowledge" that enables man for the first time to render productive people of different skills and knowledge working together in an "organization," is an innovation of this century. It has converted modern society into something brand new, something, by the way, for which we have neither political nor social theory: a society of organizations.

Books on economic history mention August Borsig as the first man to build steam locomotives in Germany. But surely far more important was his innovation—against strenuous opposition from craft guilds, teachers, and government bureaucrats—of what to this day is the German system of factory organization and the foundation of Germany's industrial strength. It was Borsig who devised the idea of the *Meister* (Master), the highly skilled and highly respected senior worker who runs the shop with considerable autonomy; and the *Lehrling System* (apprenticeship system), which combines practical training *(Lehre)* on the job with schooling *(Ausbildung)* in the classroom. And the twin inventions of modern government by Machiavelli in *The Prince* (1513) and of the modern national state by his early follower, Jean Bodin, sixty years later, have surely had more lasting impacts than most technologies.

One of the most interesting examples of social innovation and its importance can be seen in modern Japan.

From the time she opened her doors to the modern world in 1867, Japan has been consistently underrated by westerners, despite her successful defeats of China and then Russia in 1894 and 1905, respectively; despite Pearl Harbor; and despite her sudden emergence as an economic superpower and the toughest competitor in the world market of the 1970s and 1980s. A major reason, perhaps the major one, is the prevailing belief that innovation has to do with things and is based on science or technology. And the Japanese, so the common belief has held (in Japan as well as in the West, by the way), are not innovators but imitators. For the Japanese have not, by and large, produced outstanding technical or scientific innovations. Their success is based on social innovation.

When the Japanese, in the Meiji Restoration of 1867, most reluctantly opened their country to the world, it was to avoid the fates of India and nineteenth-century China, both of which were conquered, colonized, and "westernized" by the West. The basic aim, in true Judo fashion, was to use the weapons of the West to hold the West at bay; and to remain Japanese.

This meant that social innovation was far more critical than steam locomotives or the telegraph. And social innovation, in terms of the development of such institutions as schools and universities, a civil service, banks and labor relations, was far more difficult to achieve than building locomotives and telegraphs. A locomotive that will pull a train from London to Liverpool will equally, without adaptation or change, pull a train from Tokyo to Osaka. But the social institutions had to be at once quintessentially "Japanese" and yet "modern." They had to be run by Japanese and yet serve an economy that was "Western" and highly technical. Technology can be imported at low cost and with a minimum of cultural risk. Institutions, by contrast, need cultural roots to grow and to prosper. The Japanese made a deliberate decision a hundred years ago to concentrate their resources on social innovations, and to imitate, import, and adapt technical innovations—with startling success. Indeed, this policy may still be the right one for them. For, as will be discussed in Chapter 17, what is sometimes half-facetiously called creative imitation is a perfectly respectable and often very successful entrepreneurial strategy.

Even if the Japanese now have to move beyond imitating, importing, and adapting other people's technology and learn to undertake genuine technical innovation of their own, it might be prudent not to underrate them. Scientific research is in itself a fairly recent "social innovation." And the Japanese, whenever they have had to do so in the past, have always shown tremendous capacity for such innovation. Above all, they have shown a superior grasp of entrepreneurial strategies.

"Innovation," then, is an economic or social rather than a technical term. It can be defined the way J. B. Say defined entrepreneurship, as changing the yield of resources. Or, as a modern economist would tend to do, it can be defined in demand terms rather than in supply terms, that is, as changing the value and satisfaction obtained from resources by the consumer.

Which of the two is more applicable depends, I would argue, on the specific case rather than on the theoretical model. The shift from the integrated steel mill to the "mini-mill," which starts with steel scrap rather than iron ore and ends with one final product (e.g., beams and rods, rather than raw steel that then has to be fabricated), is best described and analyzed in supply terms. The end product, the end uses, and the customers are the same, though the costs are substantially lower. And the same supply definition probably fits the container. But the audiocassette or the videocassette, though equally "technical," if not more so, are better described or analyzed in terms of consumer values and consumer satisfactions, as are such social innovations as the news magazines developed by Henry Luce of Time–Life–Fortune in the 1920s, or the money-market fund of the late 1970s and early 1980s.

We cannot yet develop a theory of innovation. But we already know enough to say when, where, and how one looks systematically for innovative opportunities, and how one judges the chances for their success or the risks of their failure. We know enough to develop, though still only in outline form, the practice of innovation.

It has become almost a cliché for historians of technology that one of the great achievements of the nineteenth century was the "invention of invention." Before 1880 or so, invention was mysterious; early nineteenth-century books talk incessantly of the "flash of genius." The inventor himself was a half-romantic, half-ridiculous figure, tinkering away in a lonely garret. By 1914, the time World War I broke out, "invention" had become "research," a systematic, purposeful activity, which is planned and organized with high predictability both of the results aimed at and likely to be achieved.

Something similar now has to be done with respect to innovation. Entrepreneurs will have to learn to *practice systematic innovation.*

Successful entrepreneurs do not wait until "the Muse kisses them" and gives them a "bright idea"; they go to work. Altogether, they do not look for the "biggie," the innovation that will "revolutionize the industry," create a "billion-dollar business," or "make one rich overnight." Those entrepreneurs who start out with the idea that they'll make it big—and in a hurry—can be guaranteed failure. They are almost bound to do the wrong things. An innovation that looks very big may turn out to be nothing but technical virtuosity; and innovations with modest intellectual pretensions, a McDonald's, for instance, may turn into gigantic, highly profitable businesses. The same applies to nonbusiness, public-service innovations.

Successful entrepreneurs, whatever their individual motivation—be it money, power, curiosity, or the desire for fame and recognition—try to create value and to make a contribution. Still, successful entrepreneurs aim high. They are not content simply to improve on what already exists, or to modify it. They try to create new and different values and new and different satisfactions, to convert a "material" into a "resource," or to combine existing resources in a new and more productive configuration.

And it is change that always provides the opportunity for the new and different. *Systematic innovation therefore consists in the purposeful and organized search for changes, and in the systematic analysis of the opportunities such changes might offer for economic or social innovation.*

As a rule, these are changes that have already occurred or are under way. The overwhelming majority of successful innovations *exploit* change. To be sure, there are innovations that in themselves constitute a major change; some of the major technical innovations, such as the Wright Brothers' airplane, are examples. But these are exceptions, and fairly uncommon ones. Most successful innovations are far more prosaic; they exploit change. And thus the discipline of innovation (and it is the knowledge base of entrepreneurship) is a diagnostic discipline: a systematic examination of the areas of change that typically offer entrepreneurial opportunities.

Specifically, systematic innovation means monitoring *seven sources* for innovative opportunity.

The first four sources lie within the enterprise, whether business or public-service institution, or within an industry or service sector. They are therefore visible primarily to people within that industry or service sector. They are basically symptoms. But they are highly reliable indicators of changes that have already happened or can be made to happen with little effort. These four source areas are:

- *The unexpected*—the unexpected success, the unexpected failure, the unexpected outside event;
- *The incongruity*—between reality as it actually is and reality as it is assumed to be or as it "ought to be";
- *Innovation based on process need;*
- *Changes in industry structure or market structure* that catch everyone unawares.

The second set of sources for innovative opportunity, a set of three, involves changes outside the enterprise or industry:

- *Demographics* (population changes);
- *Changes in perception, mood, and meaning;*
- *New knowledge,* both scientific and nonscientific.

The lines between these seven source areas of innovative opportunities are blurred, and there is considerable overlap between them. They can be likened to seven windows, each on a different side of the same building. Each window shows some features that can also be seen from the window on either side of it. But the view from the center of each is distinct and different.

The seven sources require separate analysis, for each has its own distinct characteristic. No area is, however, inherently more important or more productive than the other. Major innovations are as likely to come out of an analysis of symptoms of change (such as the unexpected success of what was considered an insignificant change in product or pricing) as they are to come out of the massive application of new knowledge resulting from a great scientific breakthrough.

But the order in which these sources will be discussed is not arbitrary. They are listed in descending order of reliability and predictability. For, contrary to almost universal belief, new knowledge—and especially new scientific knowledge—is not the most reliable or most predictable source of successful innovations. For all the visibility, glamour, and importance of science-based innovation, it is actually the least reliable and least predictable one. Conversely, the mundane and unglamorous analysis of such symptoms of underlying changes as the unexpected success or the unexpected failure carry fairly low risk and uncertainty. And the innovations arising therefrom have, typically, the shortest lead time between the start of a venture and its measurable results, whether success or failure.

3

Source: The Unexpected

THE UNEXPECTED SUCCESS

No other area offers richer opportunities for successful innovation than the unexpected success. In no other area are innovative opportunities less risky and their pursuit less arduous. Yet the unexpected success is almost totally neglected; worse, managements tend actively to reject it.

Here is one example.

More than thirty years ago, I was told by the chairman of New York's largest department store, R. H. Macy, "We don't know how to stop the growth of appliance sales."

"Why do you want to stop them?" I asked, quite mystified. "Are you losing money on them?"

"On the contrary," the chairman said, "profit margins are better than on fashion goods; there are no returns, and practically no pilferage."

"Do the appliance customers keep away the fashion customers?" I asked.

"Oh, no," was the answer. "Where we used to sell appliances primarily to people who came in to buy fashions, we now sell fashions very often to people who come in to buy appliances. But," the chairman continued, "in this kind of store, it is normal and healthy for fashion to produce seventy percent of sales. Appliance sales have grown so fast that they now account for three-fifths. And that's abnormal. We've tried everything we know to make fashion grow to restore the normal ratio, but nothing works. The only thing left now is to push appliance sales down to where they should be."

For almost twenty years after this episode, Macy's New York continued to drift. Any number of explanations were given for Macy's inability to exploit its dominant position in the New York retail market: the decay of the inner city, the poor economics of a store supposedly "too big," and many others. Actually, once a new management came in after 1970, reversed the emphasis, and accepted the contribution of appliances to sales, Macy's—despite inner-city decay, despite its high labor costs, and despite its enormous size—promptly began to prosper again.

At the same time that Macy's rejected the unexpected success, another New York retail store, Bloomingdale's, used the identical unexpected success to propel itself into the number two spot in the New York market. Bloomingdale's, at best a weak number four, had been even more of a fashion store than Macy's. But when appliance sales began to climb in the early 1950s, Bloomingdale's ran with the opportunity. It realized that something unexpected was happening and analyzed it. It then built a new position in the marketplace around its Housewares Department. It also refocused its fashion and apparel sales to reach a new customer: the customer of whose emergence the explosion in appliance sales was only a symptom. Macy's is still number one in New York in volume. But Bloomingdale's has become the "smart New York store." And the stores that were the contenders for this title thirty years ago—the stores that were then strong number twos, the fashion leaders of 1950 such as Best—have disappeared (for additional examples, see Chapter 15).

The Macy's story will be called extreme. But the only uncommon aspect about it is that the chairman was aware of what he was doing. Though not conscious of their folly, far too many managements act the way Macy's did. It is never easy for a management to accept the unexpected success. It takes determination, specific policies, a willingness to look at reality, and the humility to say, "We were wrong!"

One reason why it is difficult for management to accept unexpected success is that all of us tend to believe that anything that has lasted a fair amount of time must be "normal" and go on "forever." Anything that contradicts what we have come to consider a law of nature is then rejected as unsound, unhealthy, and obviously abnormal.

This explains, for instance, why one of the major U.S. steel companies, around 1970, rejected the "mini-mill."* Management knew that

*On the "mini-mill," see Chapter 4.

its steelworks were rapidly becoming obsolete and would need billions of dollars of investment to be modernized. It also knew that it could not obtain the necessary sums. A new, smaller "mini-mill" was the solution.

Almost by accident, such a "mini-mill" was acquired. It soon began to grow rapidly and to generate cash and profits. Some of the younger men within the steel company therefore proposed that the available investment funds be used to acquire additional "mini-mills" and to build new ones. Within a few years, the "mini-mills" would then give the steel company several million tons of steel capacity based on modern technology, low labor costs, and pinpointed markets. Top management indignantly vetoed the proposal; indeed, all the men who had been connected with it found themselves "ex-employees" within a few years. "The integrated steelmaking process is the only right one," top management argued. "Everything else is cheating—a fad, unhealthy, and unlikely to endure." Needless to say, ten years later the only parts of the steel industry in America that were still healthy, growing, and reasonably prosperous were "mini-mills."

To a steelmaker who has spent his entire life working to perfect the integrated steelmaking process, who is at home in the big steel mill, and who may himself be the son of a steelworker (as a great many American steel company executives have been), anything but "big steel" is strange and alien, indeed a threat. It takes an effort to perceive in the "enemy" one's own best opportunity.

Top management people in most organizations, whether small or large, public-service institution or business, have typically grown up in one function or one area. To them, this is the area in which they feel comfortable. When I sat down with the chairman of R. H. Macy, for instance, there was only one member of top management, the personnel vice-president, who had not started as a fashion buyer and made his career in the fashion end of the business. Appliances, to these men, were something that other people dealt with.

The unexpected success can be galling. Consider the company that has worked diligently on modifying and perfecting an old product, a product that has been the "flagship" of the company for years, the product that represents "quality." At the same time, most reluctantly, the company puts through what everyone in the firm knows is a perfectly meaningless modification of an old, obsolete, and "low-quality" product. It is done only because one of the company's leading salesmen lobbied for it, or because a good customer asked for it and could not be turned down. But nobody expects it to sell; in fact, nobody wants it to sell. And then this "dog" runs away with the market and even takes the sales which plans and forecasts had promised for the "prestige," "quality" line. No wonder that everybody is appalled and considers the success a "cuckoo in the nest" (a term I have heard more than once). Everybody is likely to react precisely the way the chairman of R. H. Macy reacted when he saw the unwanted and unloved appliances overtake his beloved fashions, on which he himself had spent his working life and his energy.

The unexpected success is a challenge to management's judgment. "If the mini-mills were an opportunity, we surely would have seen it ourselves," the chairman of the big steel company is quoted as saying when he turned the mini-mill proposal down. Managements are paid for their judgment, but they are not being paid to be infallible. In fact, they are being paid to realize and admit that they have been wrong— especially when their admission opens up an opportunity. But this is by no means common.

A Swiss pharmaceutical company today has world leadership in veterinary medicines, yet it has not itself developed a single veterinary drug. But the companies that developed these medicines refused to serve the veterinary market. The medicines, mostly antibiotics, were of course developed for treating human diseases. When the veterinarians discovered that they were just as effective for animals and began to send in their orders, the original manufacturers were far from pleased. In some cases they refused to supply the veterinarians; in many others, they disliked having to reformulate the drugs for animal use, to repackage them, and so on. The medical director of a leading pharmaceutical company protested around 1953 that to apply a new antibiotic to the treatment of animals was a "misuse of a noble medicine." Consequently, when the Swiss approached this manufacturer and several others, they obtained licenses for veterinary use without any difficulty and at low cost. Some of the manufacturers were only too happy to get rid of the embarrassing success.

Human medications have since come under price pressure and are carefully scrutinized by regulatory authorities. This has made veterinary medications the most profitable segment of the pharmaceutical industry. But the companies that developed the compounds in the first place are not the ones who get these profits.

Far more often, the unexpected success is simply not seen at all. Nobody pays any attention to it. Hence, nobody exploits it, with the inevitable result that the competitor runs with it and reaps the re- wards.

A leading hospital supplier introduced a new line of instruments for biological and clinical tests. The new products were doing quite well. Then, suddenly, orders came in from industrial and university laborato- ries. Nobody was told about them, nobody noticed them; nobody real- ized that, by pure accident, the company had developed products with more and better customers outside the market for which those products had been developed. No salesman was being sent out to call on these new customers, no service force was being set up. Five or eight years later, another company had taken over these new markets. And be- cause of the volume of business these markets produced, the newcomer could soon invade the hospital market offering lower prices and better services than the original market leader.

One reason for this blindness to the unexpected success is that our existing reporting systems do not as a rule report it, let alone clamor for management's attention.

Practically every company—but every public-service institution as well—has a monthly or quarterly report. The first sheet lists the areas in which performance is below expectations: it lists the problems and the shortfalls. At the monthly meetings of the management group and the board of directors, everybody therefore focuses on the problem areas. No one even looks at the areas where the company has done better than expected. And if the unexpected success is not quantitative but qualitative—as in the case of the hospital instruments mentioned above, which opened up new major markets outside the company's traditional ones—the figures will not even show the unexpected success as a rule.

To exploit the opportunity for innovation offered by unexpected success requires analysis. Unexpected success is a symptom. But a symp- tom of what? The underlying phenomenon may be nothing more than a limitation on our own vision, knowledge, and understanding. That the pharmaceutical companies, for instance, rejected the unexpected suc- cess of their new drugs in the animal market was a symptom of their own failure to know how big—and how important—livestock raising throughout the world is; of their blindness to the sharp increase in demand for animal proteins throughout the world after World War II, and to the tremendous changes in knowledge, sophistication, and management capacity of the world's farmers.

The unexpected success of appliances at R. H. Macy's was a symptom of a fundamental change in the behavior, expectations, and values of substantial numbers of consumers—as the people at Bloomingdale's realized. Up until World War II, department store consumers in the United States bought primarily by socioeconomic status, that is, by income group. After World War II, the market increasingly segmented itself by what we now call "lifestyles." Bloomingdale's was the first of the major department stores, especially on the East Coast, to realize this, to capitalize on it, and to innovate a new retail image.

The unexpected success of laboratory instruments designed for the hospital in industrial and university laboratories was a symptom of the disappearance of distinctions between the various users of scientific instruments, which for almost a century had created sharply different markets, with different end uses, specifications, and expectations. What it symptomized—and the company never realized this—was not just that a product line had uses that were not originally envisaged. It signaled the end of the specific market niche the company had enjoyed in the hospital market. So the company that for thirty or forty years had successfully defined itself as a designer, maker, and marketer of hospital laboratory equipment was forced eventually to redefine itself as a maker of laboratory instruments, and to develop capabilities to design, manufacture, distribute, and service way beyond its original field. By then, however, it had lost a large part of the market for good.

Thus the unexpected success is not just an opportunity for innovation; it demands innovation. It forces us to ask, What basic changes are now appropriate for this organization in the way it defines its business? Its technology? Its markets? If these questions are faced up to, then the unexpected success is likely to open up the most rewarding and least risky of all innovative opportunities.

Two of the world's biggest businesses, DuPont, the world's largest chemical company, and IBM, the giant of the computer industry, owe their preeminence to their willingness to exploit the unexpected success as an innovative opportunity.

DuPont, for 130 years, had confined itself to making munitions and explosives. In the mid-1920s it then organized its first research efforts in other areas, one of them the brand-new field of polymer chemistry, which the Germans had pioneered during World War I. For several years there were no results at all. Then, in 1928, an assistant left a burner on over the weekend. On Monday morning, Wallace H. Carothers, the chemist in charge, found that the stuff in the kettle had congealed into fibers. It took another ten years before DuPont found out how to make Nylon intentionally. The point of the story is, however, that the same accident had occurred several times in the laboratories of the big German chemical companies with the same results, and much earlier. The Germans were, of course, looking for a polymerized fiber —and they could have had it, along with world leadership in the chemical industry, ten years before DuPont had Nylon. But because they had not planned the experiment, they dismissed its results, poured out the accidentally produced fibers, and started all over again.

The history of IBM equally shows what paying attention to the unexpected success can do. For IBM is largely the result of the willingness to exploit the unexpected success not once, but twice. In the early 1930s, IBM almost went under. It had spent its available money on designing the first electro-mechanical bookkeeping machine, meant for banks. But American banks did not buy new equipment in the Depression days of the early thirties. IBM even then had a policy of not laying off people, so it continued to manufacture the machines, which it had to put in storage.

When IBM was at its lowest point—so the story goes—Thomas Watson, Sr., the founder, found himself at a dinner party sitting next to a lady. When she heard his name, she said: "Are you the Mr. Watson of IBM? Why does your sales manager refuse to demonstrate your machine to me?" What a lady would want with an accounting machine Thomas Watson could not possibly figure out, nor did it help him much when she told him she was the director of the New York Public Library; it turned out he had never been in a public library. But next morning, he appeared there as soon as its doors opened.

In those days, libraries had fair amounts of government money. Watson walked out two hours later with enough of an order to cover next month's payroll. And, as he added with a chuckle whenever he told the story, "I invented a new policy on the spot: we get cash in advance before we deliver."

Fifteen years later, IBM had one of the early computers. Like the other early American computers, the IBM computer was designed for scientific purposes only. Indeed, IBM got into computer work largely because of Watson's interest in astronomy. And when first demonstrated in IBM's show window on Madison Avenue, where it drew enormous crowds, IBM's computer was programmed to calculate all past, present, and future phases of the moon.

But then businesses began to buy this "scientific marvel" for the most mundane of purposes, such as payroll. Univac, which had the most advanced computer and the one most suitable for business uses, did not really want to "demean" its scientific miracle by supplying business. But IBM, though equally surprised by the business demand for computers, responded immediately. Indeed, it was willing to sacrifice its own computer design, which was not particularly suitable for accounting, and instead use what its rival and competitor (Univac) had developed. Within four years IBM had attained leadership in the computer market, even though for another decade its own computers were technically inferior to those produced by Univac. IBM was willing to satisfy business and to satisfy it on business' terms—to train programmers for business, for instance.

Similarly, Japan's foremost electronic company, Matsushita (better known by its brand names Panasonic and National), owes its rise to its willingness to run with unexpected success.

Matsushita was a fairly small and undistinguished company in the early 1950s, outranked on every count by such older and deeply entrenched giants as Toshiba or Hitachi. Matsushita "knew," as did every other Japanese manufacturer of the time, that "television would not grow fast in Japan." "Japan is much too poor to afford such a luxury," the chairman of Toshiba had said at a New York meeting around 1954 or 1955. Matsushita, however, was intelligent enough to accept that the Japanese farmers apparently did not know that they were too poor for television. What they knew was that television offered them, for the first time, access to a big world. They could not afford television sets, but they were prepared to buy them anyhow and pay for them. Toshiba and Hitachi made better sets at the time, only they showed them on the Ginza in Tokyo and in the big-city department stores, making it pretty clear that farmers were not particularly welcome in such elegant surroundings. Matsushita went to the farmers and sold its televisions door-to-door, something no one in Japan had ever done before for anything more expensive than cotton pants or aprons.

Of course, it is not enough to depend on accidents, nor to wait for the lady at the dinner table to express unexpected interest in one's apparently failing product. The search has to be organized.

The first thing is to ensure that the unexpected is being seen; indeed, that it clamors for attention. It must be properly featured in the information management obtains and studies. (How to do this is described in some detail in Chapter 13.)

Managements must look at every unexpected success with the questions: (1) What would it mean to us if we exploited it? (2) Where could it lead us? (3) What would we have to do to convert it into an opportunity? And (4) How do we go about it? This means, first, that managements need to set aside specific time in which to discuss unexpected successes; and second, that someone should always be designated to analyze an unexpected success and to think through how it could be exploited.

But management also needs to learn what the unexpected success demands of them. Again, this might best be explained by an example.

A major university on the eastern seaboard of the United States started, in the early 1950s, an evening program of "continuing education" for adults, in which the normal undergraduate curriculum leading to an undergraduate degree was offered to adults with a high school diploma.

Nobody on the faculty really believed in the program. The only reason it was offered at all was that a small number of returning World War II veterans had been forced to go to work before obtaining their undergraduate degrees and were clamoring for an opportunity to get the credits they still lacked. To everybody's surprise, however, the program proved immensely successful, with qualified students applying in large numbers. And the students in the program actually outperformed the regular undergraduates. This, in turn, created a dilemma. To exploit the unexpected success, the university would have had to build a fairly big first-rate faculty. But this would have weakened its main program; at the least, it would have diverted the university from what it saw as its main mission, the training of undergraduates. The alternative was to close down the new program. Either decision would have been a responsible one. Instead, the university decided to staff the program with cheap, temporary faculty, mostly teaching assistants working on their own advanced degrees. As a result, it destroyed the program within a few years; but worse, it also seriously damaged its own reputation.

The unexpected success is an opportunity, but it makes demands. It demands to be taken seriously. It demands to be staffed with the ablest people available, rather than with whoever we can spare. It demands seriousness and support on the part of management equal to the size of the opportunity. And the opportunity is considerable.

II

THE UNEXPECTED FAILURE

Failures, unlike successes, cannot be rejected and rarely go unnoticed. But they are seldom seen as symptoms of opportunity. A good many failures are, of course, nothing but mistakes, the results of greed, stupidity, thoughtless bandwagon-climbing, or incompetence whether in design or execution. Yet if something fails despite being carefully planned, carefully designed, and conscientiously executed, that failure often bespeaks underlying change and, with it, opportunity.

The assumptions on which a product or service, its design or its marketing strategy, were based may no longer fit reality. Perhaps customers have changed their values and perceptions; while they still buy the same "thing," they are actually purchasing a very different "value." Or perhaps what has always been one market or one end use is splitting itself into two or more, each demanding something quite different. Any change like this is an opportunity for innovation.

I had my first experience with an unexpected failure at the very beginning of my working life, almost sixty years ago, just out of high school. My first job was as a trainee in an old export firm, which for more than a century had been selling hardware to British India. Its best seller for years had been a cheap padlock, of which it exported whole shiploads every month. The padlock was flimsy; a pin easily opened the lock. As incomes in India went up during the 1920s, padlock sales, instead of going up, began to decline quite sharply. My employer thereupon did the obvious: he redesigned the padlock to give it a sturdier lock, that is, to make it "better quality." The added cost was minimal and the improvement in quality substantial. But the improved padlock turned out to be unsalable. Four years later, the firm went into liquidation, the decline of its Indian padlock business a major factor in its demise.

A very small competitor of this firm in the Indian export business—no more than a tenth of the size of my employer and until then barely able to survive—realized that this unexpected failure was a symptom of basic change. For the bulk of Indians, the peasants in the villages, the padlock was (and for all I know, still is) a magical symbol; no thief would have dared open a padlock. The key was never used, and usually disappeared. To get a padlock that could not easily be opened without a key —the improved padlock my employer had worked so hard to perfect without additional cost—was thus not a boon but a disaster.

A small but rapidly growing middle-class minority in the cities, however, needed a real lock. That it was not sturdy enough for their needs was the main reason why the old lock had begun to lose sales and market. But for them the redesigned product was still inadequate.

My employer's competitor broke down the padlock into two separate products: one without lock and key, with only a simple trigger release, and selling for one-third less than the old padlock but with twice its profit margin; and the other with a good sturdy lock and three keys, selling at twice the price of the old product and also with a substantially larger profit margin. Both lines immediately began to sell. Within two years, the competitor had become the largest European hardware exporter to India. He maintained this position for ten years, until World War II put an end to European exports to India altogether.

A quaint tale from horse and buggy days, some might say. Surely we have become more sophisticated in this age of computers, of market research, and of business school MBAs.

But here is another case, half a century later and from a very "sophisticated" industry. Yet it teaches exactly the same lesson.

Just at the time when the first cohorts of the "baby boom" were reaching their mid-twenties—that is, the age to form families and to buy their first house—the 1973–74 recession hit. Inflation was becoming rampant, particularly in housing prices, which rose much faster than anything else. At the same time, interest rates on home mortgages were skyrocketing. And so the mass builders in America began to design and offer what they called a "basic house," smaller, simpler, and cheaper than the house that had become standard.

But despite its being such "good value" and well within the means of the first-time homebuyer, the "basic house" was a thumping failure. The builders tried to salvage it by offering low-interest financing and long repayment terms, and by slashing prices. Still, no one bought the "basic house."

Most homebuilders did what businessmen do in an unexpected failure: they blamed that old bogeyman, the "irrational customer." But one builder, still very small, decided to look around. He found that there had been a change in what the young American couple wants in its first house. This no longer represents the family's permanent home as it had done for their grandparents, a house in which the couple expects to live the rest of its life, or at least a long time. In the 1970s, young couples were buying not one, but two separate "values" in purchasing their first home. They bought shelter for a few short years; and they also bought an option to buy—a few years later—their "real" house, a much bigger and more luxurious home, in a better neighborhood, with better schools. To make the down payment on this far more expensive permanent home, they would, however, need the equity they had built up in the first house. The young people knew very well that the "basic house" was not what they and their contemporaries really wanted, even though it was all they could afford. They feared therefore—and perfectly rationally—that they would not be able to resell the "basic house" at a decent price. So the "basic house," instead of being an option to buy the "real house" later on, would become a serious impediment to the fulfillment of their true housing needs and wants.

The young couple of 1950 had still perceived itself as "working-class," by and large. And "working-class" people in the West do not expect their incomes and their standards of living to rise materially once they are out of their apprenticeship and into a full-time job. Seniority, for working-class people (with Japan being the major exception), means greater job security rather than larger incomes. But the "middle class" traditionally can expect a steady increase in its income until the head of the household reaches age forty-five or forty-eight. Between 1950 and 1975, both the reality and the self-image of young American adults—their educations, their expectations, their jobs—had changed from "working-class" to "middle-class." And with this change had come a sharp change in what the young people's first home represented, and what "value" was connected with it.

Once this was understood—and all it took was to listen to prospective homebuyers for a few weekends—successful innovation came about easily. Almost no change was made in the physical plant itself; only the kitchen was redesigned and made somewhat roomier. Otherwise, the building remained the same "basic house" the homebuilders had not been able to sell. But instead of being offered as "your house," it was offered as "your *first* house," and as a "building block toward the house you want." Specifically, this meant that the young couple was

shown both the house as it was standing—that is, the "basic house"—
and a model of the same house in which future additions such as an
extra bathroom, one or two more bedrooms, and a basement "family
den" had been built. Indeed, the builder had already obtained the
necessary city permits for conversion of the "basic house" to a "permanent home." Furthermore, the builder guaranteed the young couple a
fixed resale price for their first house, to be credited against their purchase from his firm of a second, bigger, "permanent" home within five
to seven years. "This entailed practically no risk," he explained. "The
demographics were such, after all, as to guarantee a steady increase in
the demand for 'first houses' until the late 1980s or 1990s, during which
time the babies of the 'baby bust' of 1961 will have become twenty-five
themselves and will start forming their own families."

Before this homebuilder transformed failure into innovation, he
had operated in only one metropolitan area and was a small factor in
it. Five years later, the firm was operating in seven metropolitan areas
and was either number one or a strong number two in each of them.
Even during the building recession of 1981–82—a recession so severe
that some of the largest American builders did not sell one single new
house during an entire season—this innovative homebuilder continued to grow. "One reason," the firm's founder explained, "was
something even I had not seen when I decided to offer first-time
homebuyers a repurchase guarantee. It gave us a steady supply of
well-built and still fairly new houses that needed only a little fixing up
and could then be resold at a very decent profit to the next crop of
first-home buyers."

Faced with unexpected failure, executives, especially in large organizations, tend to call for more study and more analysis. But as both the
padlock story and the "basic house" story show, this is the wrong response. The unexpected failure demands that you go out, look around,
and listen. Failure should always be considered a symptom of an innovative opportunity, and taken seriously as such.

It is equally important to watch out for the unexpected event in a
supplier's business, and among the customers. McDonald's, for instance,
started because the company's founder, Ray Kroc, paid attention to the
unexpected success of one of his customers. At that time Kroc was
selling milkshake machines to hamburger joints. He noticed that one of
his customers, a small hamburger stand in a remote California town,
bought several times the number of milkshake machines its location

and size could justify. He investigated and found an old man who had, in effect, reinvented the fast-food business by systematizing it. Kroc bought his outfit and built it into a billion-dollar business based on the original owner's unexpected success.

A competitor's unexpected success or failure is equally important. In either case, one takes the event seriously as a possible symptom of innovative opportunity. One does not just "analyze." One goes out to investigate.

Innovation—and this is a main thesis of this book—is organized, systematic, rational work. But it is perceptual fully as much as conceptual. To be sure, what the innovator sees and learns has to be subjected to rigorous logical analysis. Intuition is not good enough; indeed, it is no good at all if by "intuition" is meant "what I feel." For that usually is another way of saying "What I like it to be" rather than "What I perceive it to be." But the analysis, with all its rigor—its requirements for testing, piloting, and evaluating—has to be based on a perception of change, of opportunity, of the new realities, of the incongruity between what most people still are quite sure is the reality and what has actually become a new reality. This requires the willingness to say: "I don't know enough to analyze, but I shall find out. I'll go out, look around, ask questions, and *listen.*"

It is precisely because the unexpected jolts us out of our preconceived notions, our assumptions, our certainties, that it is such a fertile source of innovation.

It is not in fact even necessary for the entrepreneur to understand why reality has changed. In the two cases above, it was easy to find out what had happened and why. More often, we find out what is happening without much clue as to why. And yet we can still innovate successfully.

Here is one example.

The failure of the Ford Motor Company's Edsel in 1957 has become American folklore. Even people who were not yet born when the Edsel failed have heard about it, at least in the United States. But the general belief that the Edsel was a slapdash gamble is totally mistaken.

Very few products were ever more carefully designed, more carefully introduced, more skillfully marketed. The Edsel was intended to be the final step in the most thoroughly planned strategy in American business history: a ten-year campaign during which the Ford Motor Company converted itself after World War II from near-bankruptcy into an aggressive competitor, a strong number two in the United States, and a few years later, a strong contender for the number one spot in the rapidly growing European market.

By 1957, Ford had already successfully reestablished itself as a strong competitor in three of the four main American automobile markets: the "standard" one with the Ford nameplate; the "lower-middle" one with Mercury; and the "upper" one with the Continental. The Edsel was then designed for the only remaining segment, the upper-middle one, the one for which Ford's big rival, General Motors, produced the Buick and the Oldsmobile. This "upper-middle" segment was, in the period after World War II, the fastest-growing part of the automobile market and yet the one for which the third automobile producer, Chrysler, did not have a strong entry, thereby leaving the door wide open for Ford.

Ford went to extreme lengths to plan and design the Edsel, embodying in its design the best information from market research, the best information about customer preferences in appearance and styling, and the highest standards of quality control.

Yet the Edsel became a total failure right away.

The reaction of the Ford Motor Company was very revealing. Instead of blaming the "irrational consumer," the Ford people decided there was something happening that did not jibe with the assumptions about reality everyone in the automobile industry had been making about consumer behavior—and for so long that they had become unquestioned axioms.

The result of Ford's decision to go out and investigate was the one genuine innovation in the American automobile industry since Alfred P. Sloan, in the 1920s, had defined the socioeconomic segmentation of the American market into "low," "lower-middle," "upper-middle," and "upper" segments, the insight on which he then built the General Motors Company. When the Ford people went out, they discovered that this segmentation was rapidly being replaced—or at least paralleled—by another quite different one, the one we would now call "lifestyle segmentation." The result, within a short period after the Edsel's failure, was the appearance of Ford's Thunderbird, the greatest success of any American car since Henry Ford, Sr., had introduced his Model T in 1908. The Thunderbird established Ford again as a major producer in its own right, rather than as GM's kid brother and a perennial imitator.

And yet to this day we really do not know what caused the change. It occurred well *before* any of the events by which it is usually explained, such as the shift of the center of demographic gravity to the teenagers as a result of the "baby boom," the explosive expansion of higher education, or the change in sexual mores. Nor do we really know what is meant by "lifestyle." All attempts to describe it have been futile so far. All we know is that something happened.

But that is enough to convert the unexpected, whether success or failure, into an opportunity for effective and purposeful innovation.

III

Unexpected successes and unexpected failure have so far been discussed as occurring within a business or an industry. But outside events, that is, events that are not recorded in the information and the figures by which a management steers its institution, are just as important. Indeed, they often are more important.

Here are some examples showing typical unexpected outside events and their exploitation as major opportunities for successful innovation.

One example concerns IBM and the personal computer.

However much executives and engineers at IBM may have disagreed with each other, there apparently was total agreement within the company on one point until well into the seventies: the future belonged to the centralized "main-frame" computer, with an ever larger memory and an ever larger calculating capacity. Everything else, every IBM engineer could prove convincingly, would be far too expensive, far too confusing, and far too limited in its performance capacity. And so IBM concentrated its efforts and resources on maintaining its leadership in the main-frame market.

And then around 1975 or 1976, to everybody's total surprise, ten- and eleven-year-old kids began to play computer games. Right away their fathers wanted their own office computer or personal computer, that is, a separate, small, freestanding machine with far less capacity than even the smallest main-frame has. All the dire things the IBM people had predicted actually did happen. The freestanding machines cost many times what a plug-in "terminal" costs, and they have far less capacity; there is such a proliferation of them and their programs, and so few of them are truly compatible with one another, that the whole field has become chaotic, with service and repairs in shambles. But this does not seem to bother the customers. On the contrary, in the U.S. market the personal computers in five short years—from 1979 to 1984 —reached the annual sales volume it had taken the "main-frames" thirty years to reach, that is, $15–$16 billion.

IBM could have been expected to dismiss this development. Instead, as early as 1977, when personal computer sales worldwide were still less than $200 million (as against main-frame sales of $7 billion for the same year), IBM set up task forces in competition with one another to develop personal computers for the company. As a result, IBM produced its own personal computer in 1980, just when the market was exploding. Three years later, in 1983, IBM had become the world's leading personal computer producer with nearly as much of a leadership position in the new field as it had in main-frames. Also in 1983 IBM then introduced its own very small "home computer," the "Peanut."

When I discuss all this with the IBM people, I always ask the same question: "What explains that IBM, of all people, saw this change as an opportunity when everybody at IBM was so totally sure that it couldn't happen and made no sense?" And I always get the same answer: "Precisely because we *knew* that this couldn't happen, and that it would make no sense at all, the development came as a profound shock to us. We realized that everything we'd assumed, everything we were so absolutely certain of, was suddenly being thrown into a cocked hat, and that we had to go out and organize ourselves to take advantage of a development we knew couldn't happen, but which then did happen."

The second example is far more mundane. But is it no less instructive despite its lack of glamour.

The United States has never been a book-*buying* country, in part because of the ubiquitous free public library. When TV appeared in the early fifties and more and more Americans began to spend more and more of their time in front of the tube—particularly people in their prime book-reading years, that is, people of high school and college age —"everyone knew" that book sales would drop drastically. Book publishers frantically began to diversify into "high-tech media": educational movies, or computer programs (in most cases, with total lack of success). But instead of collapsing, book sales in the United States have soared since TV first came in. They have grown several times as fast as every indicator had predicted, whether family incomes, total population in the "book-reading years," or even people with higher degrees.

No one knows why this happened. Indeed, no one quite knows what really happened. Books are still as rare in the typical American home as before.* Where, then, do all these books go? That we have no answer to this question does not alter the fact that books are being bought and paid for in increasing numbers.

Both the publishers and the existing bookstores knew, of course, all along that book sales were soaring. Neither, however, did anything about it. The unexpected event was exploited, instead, by a few mass retailers such as department stores in Minneapolis and Los Angeles. None of these people had ever had anything to do with books, but they knew the retail business. They started bookstore chains that are quite different from any earlier bookstore in America. Basically, these are supermarkets. They do not treat books as literature but as "mass merchandise," and they concentrate on the fast-moving items that generate the largest dollar sales per unit of shelf space. They are located in shopping centers with high rents but also with high traffic, whereas everybody in the book business had known all along that a bookstore has to be in a low-rent location, preferably near a university. Traditionally, booksellers were themselves "literary types" and tried to hire people who "love books." The managers of the new bookstores are former cosmetics salespeople. The standing joke among them is that any salesperson who wants to read anything besides the price tag on the book is hopelessly overqualified.

For ten years now, these new bookstore chains have been among the most successful and fastest-growing segments in American retailing and among the fastest-growing new businesses in this country altogether.

Each of these cases represents genuine innovation. But not one of them represents diversification.

IBM stayed in the computer business. And the chain bookstores are run by people who all along have been in retailing, in shopping centers, or managing "boutiques."

It is a condition of success in exploiting the unexpected outside event that it must fit the knowledge and expertise of one's own business. Companies, even large companies, that went into the new book market

---

*This is also true of Japan, the country that, *per capita*, buys more books than any other and twice as many as the United States.

or into mass merchandising without the retail expertise have uniformly come to grief.

The unexpected outside event may thus be, above all, an opportunity to apply already existing expertise to a new application, but to an application that does not change the nature of the "business we are in." It may be extension rather than diversification. Yet as the above examples show, it also demands innovation in product and often in service and distribution channels.

The second point about these cases is that they all are big-company cases. Of course, a good many of the cases in this book, as in any management book, have to be big-company cases. They are the only available ones, as a rule, the only ones that can be found in the published records, the only ones discussed on the business page of newspapers or in magazines. Small-company cases are much harder to come by and often cannot be discussed without violating confidences.

But exploiting the unexpected outside event appears to be something that particularly fits the existing enterprise, and a fairly sizable one at that. I know of few small companies that have successfully exploited the unexpected outside event; nor does any other student of entrepreneurship and innovation whom I could consult. This may be coincidence. But perhaps the existing large enterprise is more likely to see the "big picture."

It is the large retailer in the United States who is used to looking at figures that show where and how consumers spend retail dollars. The large retailer also knows about shopping-center locations and how to get the good ones. And could a small company have done what IBM did and detach four task forces of first-rate designers and engineers to work on new product lines? Smaller high-tech companies in a rapidly growing industry usually do not have enough of such people even for their existing work.

It may well be that the unexpected outside event is the innovative area that offers the large enterprise the greatest opportunity along with the lowest risk. It may be the area that is particularly suited for innovation by the large and established enterprise. It may be the area in which expertise matters the most, and in which the ability to mobilize substantial resources fast makes the greatest difference.

But as these cases also show, being big and established does not guarantee that an enterprise will perceive the unexpected event and successfully organize itself to exploit it. IBM's American competitors are all big businesses with sales in the billions. Not one of them exploited the personal computer—they were all too busy fighting IBM. And not one of the old large bookstore chains in the United States, Brentano's in New York, for instance, exploited the new book market.

The opportunity is there, in other words. It is a major opportunity, occurring frequently. And when it occurs, it holds out great promise, particularly for existing and sizable enterprises. But such opportunities require more than mere luck or intuition. They demand that the enterprise search for innovation, be organized for it, and be managed so as to exploit it.

4

Source: Incongruities

An incongruity is a discrepancy, a dissonance, between what is and what "ought" to be, or between what is and what everybody assumes it to be. We may not understand the reason for it; indeed, we often cannot figure it out. Still, an incongruity is a symptom of an opportunity to innovate. It bespeaks an underlying "fault," to use the geologist's term. Such a fault is an invitation to innovate. It creates an instability in which quite minor efforts can move large masses and bring about a restructuring of the economic or social configuration. Incongruities do not, however, usually manifest themselves in the figures or reports executives receive and pay attention to. They are qualitative rather than quantitative.

Like the unexpected event, whether success or failure, incongruity is a symptom of change, either change that has already occurred or change that can be made to happen. Like the changes that underlie the unexpected event, the changes that underlie incongruity are changes *within* an industry, a market, a process. The incongruity is thus clearly visible to the people within or close to the industry, market, or process; it is directly in front of their eyes. Yet it is often overlooked by the insiders, who tend to take it for granted—"This is the way it's always been," they say, even though "always" may be a very recent development.

There are several kinds of incongruity:

— An incongruity between the economic realities of an industry (or of a public-service area);

— An incongruity between the reality of an industry (or of a public-service area) and the assumptions about it;

— An incongruity between the efforts of an industry (or a public-

service area) and the values and expectations of its customers;
— An internal incongruity within the rhythm or the logic of a process.

I

INCONGRUOUS ECONOMIC REALITIES

If the demand for a product or a service is growing steadily, its economic performance should steadily improve, too. It should be easy to be profitable in an industry with steadily rising demand. The tide carries it. A lack of profitability and results in such an industry bespeaks an incongruity between economic realities.

Typically, these incongruities are macro-phenomena, which occur within a whole industry or a whole service sector. The major opportunities for innovation exist, however, normally for the small and highly focused new enterprise, new process, or new service. And usually the innovator who exploits this incongruity can count on being left alone for a long time before the existing businesses or suppliers wake up to the fact that they have new and dangerous competition. For they are so busy trying to bridge the gap between rising demand and lagging results that they barely even notice somebody is doing something different—something that produces results, that exploits the rising demand.

Sometimes we understand what is going on. But sometimes it is impossible to figure out why rising demand does not result in better performance. The innovator, therefore, need not always try to understand why things do not work as they should. He should ask instead: "What would exploit this incongruity? What would convert it into an opportunity? What can be done?" Incongruity between economic realities is a call to action. Sometimes the action to be taken is rather obvious, even though the problem itself is quite obscure. And sometimes we understand the problem thoroughly and yet cannot figure out what to do about it.

The steel "mini-mill" is a good example of an innovation that successfully exploited incongruity.

For more than fifty years, since the end of World War I, the large, integrated steel mill in developed countries did well only in wartime. In times of peace its results were consistently disappointing, even though the demand for steel appeared to be going up steadily, at least until 1973.

The explanation of this incongruity has long been known. The minimum incremental unit needed to satisfy additional demand in an integrated steel mill is a very big investment and adds substantially to capacity. Any expansion to an existing steel mill is thus likely to operate for a good many years at a low utilization rate, until demand—which always goes up in small, incremental steps except in wartime—reaches the new capacity level. But not to expand when demand creeps up means losing market share, and permanently. No company can afford to take that risk. The industry can therefore only be profitable for a few short years: between the time when everybody begins to build new capacity and the time when all this new capacity comes on stream.

Further, the steelmaking process invented in the 1870s is fundamentally uneconomical, as also has been known for many years. It tries to defy the laws of physics—and that means violating the laws of economics. Nothing in physics requires as much work as the creation of temperatures, whether hot or cold, unless it is working against the laws of gravity and of inertia. The integrated steel process creates very high temperatures four times, only to quench them again. And it lifts heavy masses of hot materials and then moves them over considerable distances.

It had been clear for many years that the first innovation in process that would assuage these inherent weaknesses would substantially lower costs. This is exactly what the "mini-mill" does. A mini-mill is not a "small" plant; the minimum economical size produces around $100 million of sales. But that is still about one-sixth to one-tenth the minimum economic size of an integrated steel mill. A mini-mill can thus be built to provide, economically, a fairly small additional increment of steel production for which the market already exists. The mini-mill creates heat only once, and does not quench it, but uses it for the rest of the process. It starts with steel scrap instead of iron ore, and then concentrates on one end product: sheet, for instance, or beams, or rods. And while the integrated steel mill is highly labor-intensive, the mini-mill can be automated. Its costs thus come to less than half those of the traditional steel process.

Governments, labor unions, and the integrated steel companies have been fighting the mini-mill every step of the way. But it is steadily encroaching. By the year 2000, fifty percent or more of the steel used in the United States is likely to come out of mini-mills, while the large, integrated steel mills will be in irreversible decline.

There is a catch, however, and it is an important one. A similar incongruity between the economic reality of demand and the economic reality of the process exists in the paper industry. Only in this case, we do not know how to convert it into innovation and opportunity.

Despite the constant efforts of the governments of all developed and most developing countries to increase the demand for paper—perhaps the only objective on which the governments of all countries agree— the paper industry has not been doing well. Three years of "record profits" are invariably followed by five years of "excess capacity" and losses. Yet we do not, so far, have anything like a "mini-mill" process for paper. For eighty or ninety years, it has been known that wood fiber is a monomer; and it should not be too difficult, one would say, to find a plasticizer that converts it into a polymer. This would convert paper-making from an inherently inefficient and wasteful mechanical process into an inherently efficient chemical process. Indeed, almost a hundred years ago this was achieved as far as making textile fibers out of wood pulp is concerned—in the rayon process, which dates back to the 1880s. But despite millions spent in research, nobody has so far found a technique to produce paper that way.

In an incongruity, as these cases exemplify, the innovative solution has to be clearly definable. It has to be feasible with the existing, known technology, and with easily available resources. It requires hard developmental work, of course. But if a great deal of research and new knowledge is still needed, it is not yet ready for the entrepreneur, not yet "ripe." The innovation that successfully exploits an incongruity between economic realities has to be simple rather than complicated, "obvious" rather than grandiose.

In public-service areas, too, major incongruities between economic realities can be found.

Health care in developed countries offers one example. As recently as 1929, health care represented an insignificant portion of national expenditure in all developed countries, taking up a good deal less than 1 percent of gross national product or of consumer expenditures. Now, half a century later, health care, and expecially the hospital, accounts in all developed countries for 7 to 11 percent of a much larger gross national product. Yet economic performance has been going down rather than up. Costs have risen much faster than services—perhaps three or four times as fast. The demand will continue to rise with the steady growth in the number of older people in all developed countries over the next thirty years. And so will the costs, which are closely tied to the age of the population.

We do not understand the phenomenon.* But successful innovations, simple, targeted and focused on specific objectives, have emerged in Great Britain and the United States. These innovations are quite different simply because the two countries have such radically different systems. But each exploits the specific vulnerability of its country's system and converts it into an opportunity.

In Britain, the "radical innovation" is private health insurance, which has become the fastest-growing and most popular employee benefit. All it does is to enable policyholders to be seen immediately by a specialist and to jump to the head of the queue and avoid having to wait should they need "elective surgery."† For the British system has attempted to keep health-care costs down by *"triage"* which, in effect, reserves immediate attention and treatment to routine illnesses on the one hand and to "life-threatening" ailments on the other, but puts everything else, and especially elective surgery, on hold with waiting periods now running into years (e.g., for replacing a hip destroyed by arthritis). Health insurance policyholders, however, are operated on right away.

In contrast to Great Britain, the United States has so far tried to satisfy all demands of health care regardless of cost. As a result, hospital costs in America have exploded. This created a different innovative opportunity: to "unbundle," that is, to move out of the hospital into separate locations a host of services that do not require such high-cost hospital facilities as a body scanner or cobalt X-Ray to treat cancers, the highly instrumented and automated medical laboratory, or physical rehabilitation. Each of these innovative responses is small and specific: a freestanding maternity center, which basically offers motel facilities for mother and new baby; a freestanding "ambulatory" surgical center for surgery that does not require a hospital stay and post-operative care;

---

*This is brought out clearly in the best discussion of the health-care problem that has appeared so far, and the only one that looks at health care across national boundaries, in all developed countries. It is given in *The Economist* of April 29, 1984.

†Surgery for complaints that yield to surgery, will not improve without it, but are not "life-threatening." Examples are cataracts, hip replacements and orthopedic surgery generally, or a prolapsed uterus.

a psychiatric diagnostic and referral center; geriatric centers of a similar nature; and so on.

These new facilities do not substitute for the hospital. What they do in effect is to push the American hospital toward the same role the British have assigned to their hospitals: as a place for emergencies, for life-threatening diseases, and for intensive and acute sickness care. But these innovations which, as in Britain, are embodied primarily in profit-making "businesses," convert the incongruity between the economic reality of rising health-care demand and the economic reality of falling health-care performance into an opportunity for innovation.

These are "big" examples, taken from major industries and public services. It is this fact, however, that makes them accessible, visible, and understandable. Above all, these examples show why the incongruity between economic realities offers such great innovative opportunities. The people who work within these industries or public services know that there are basic flaws. But they are almost forced to ignore them and to concentrate instead on patching here, improving there, fighting this fire or caulking that crack. They are thus unable to take the innovation seriously, let alone to try to compete with it. They do not, as a rule, even notice it until it has grown so big as to encroach on their industry or service, by which time it has become irreversible. In the meantime, the innovators have the field to themselves.

II

THE INCONGRUITY BETWEEN REALITY
AND THE ASSUMPTIONS ABOUT IT

Whenever the people in an industry or a service misconceive reality, whenever they therefore make erroneous assumptions about it, their efforts will be misdirected. They will concentrate on the area where results do not exist. Then there is an incongruity between reality and behavior, an incongruity that once again offers opportunity for successful innovation to whoever can perceive and exploit it.

A simple example is that old workhorse of world trade, the ocean-going general cargo vessel.

Thirty-five years ago, in the early 1950s, the ocean-going freighter was believed to be dying. The general forecast was that it would be replaced by air freight, except for bulk commodities. Costs of ocean freight were rising at a fast clip, and it took longer and longer to get merchandise delivered by freighter as one port after another became badly congested. This, in turn, increased pilferage at the docks as more and more merchandise piled up waiting to be loaded while vessels could not make it to the pier.

The basic reason was that the shipping industry had misdirected its efforts toward nonresults for many years. It had tried to design and build faster ships, and ships that required less fuel and a smaller crew. It concentrated on the economics of the ship while at sea and in transit from one port to another.

But a ship is capital equipment; and for all capital equipment the biggest cost is the cost of not working, during which interest has to be paid while the equipment does not earn. Everybody in the industry knew, of course, that the main expense of a ship is interest on the investment. Yet the industry kept on concentrating its efforts on costs that were already quite low—the costs of the ship while at sea and doing work.

The solution was simple: Uncouple loading from stowing. Do the loading on land, where there is ample space and where it can be performed before the ship is in port, so that all that has to be done is to put on and take off pre-loaded freight. Concentrate, in other words, on the costs of not working rather than on those of working. The answer was the roll-on, roll-off ship and the container ship.

The results of these simple innovations have been startling. Freighter traffic in the last thirty years has increased up to five-fold. Costs, overall, are down by 60 percent. Port time has been cut by three-quarters in many cases, and with it congestion and pilferage.

Incongruity between perceived reality and actual reality often declares itself. But whenever serious, concentrated efforts do not make things better but, on the contrary, make things worse—where faster ships only mean more port congestion and longer delivery times—it is highly probable that efforts are being misdirected. In all likelihood, refocusing on where the results are will yield substantial returns easily and fast.

Indeed, the incongruity between perceived and actual reality rarely requires "heroic" innovations. Uncoupling the loading of freight from the stowing thereof required little but adapting to the ocean-going freighter methods which, much earlier, had been developed for trucks and railroads.

The incongruity between perceived and actual reality typically characterizes a whole industry or a whole service area. The solution, however, should again be small and simple, focused and highly specific.

III

THE INCONGRUITY BETWEEN
PERCEIVED AND ACTUAL CUSTOMER
VALUES AND EXPECTATIONS

In Chapter 3, I mentioned the case of television in Japan as an example of the unexpected success. It is also a good example of the incongruity between actual and perceived customer values and customer expectations. Long before the Japanese industrialist told his American audience that the poor in his country would not buy a TV set because they could not afford it, the poor in the United States and in Europe had already shown that TV satisfies expectations which have little to do with traditional economics. But this highly intelligent Japanese simply could not conceive that for customers—and especially for poor customers—the TV set is not just a "thing." It represents access to a new world; access, perhaps, to a whole new life.

Similarly, Khrushchev could not conceive that the automobile is not a "thing" when he said on his visit to the United States in 1956 that "Russians will never want to own automobiles; cheap taxis make much more sense." Any teenager could have told him that "wheels" are not mere transportation but freedom, mobility, power, romance. And Khrushchev's misperception created one of the wildest entrepreneurial opportunities: the shortage of automobiles in Russia has brought forth the biggest and liveliest black market.

These, it will be said, are again "cosmic" examples, not much use to a businessman or to an executive in a hospital, a university, or a trade association. But they are examples of a common phenomenon. What follows is a different case, in its own way equally "cosmic" but very definitely of operational significance.

One of the fastest-growing American financial institutions for the last several years has been a securities firm located not in New York but in a suburb of a Midwestern city. It now has two thousand branch offices all over the United States. And it owes its success and growth to having exploited an incongruity.

The large financial institutions, the Merrill Lynches and Dean Witters and E. F. Huttons, assume that their customers have the same values they have. To them it is obvious, if not axiomatic, that people invest in order to get rich. This is, after all, what motivates the members of the New York Stock Exchange, and determines what they consider "success." However, this assumption holds true only for a part of the investing public, and surely not even for the majority. They are not "financial people." They know that in order to "get rich" by investing, one has to work full time at managing money and be pretty knowledgeable about it. The local professional men, the local small businessmen, the local substantial farmers, however, have neither such time nor such knowledge; they are much too busy earning their money to have time to manage it.

This is the incongruity which the Midwestern securities firm exploits. Outwardly, it looks just like any other securities firm. It is a member of the New York Stock Exchange. But only a very small portion of its business, around one-eighth, is Stock Exchange business. It stays away from the items the big trading houses on Wall Street push the hardest: options, commodity futures, and so on, appealing instead to what it calls "the intelligent investor." It does not promise—and this is a genuine innovation among American financial service institutions—that its customers will make a fortune. It does not even want customers who trade. It wants customers who earn more money than they spend, which is typical for the successful professional, the substantial farmer, or the small-town businessman, less because their incomes are high than because their spending habits are modest. And then it appeals to their psychological need to protect their money. What this firm sells is a chance to maintain one's savings—through investment in bonds and stocks, to be sure, but also in deferred annuities, tax-sheltered partnerships, real estate trust, and so on. The "product" the firm delivers is a different one and one that no Wall Street house has ever sold before: peace of mind. And this is what really represents "value" for the "intelligent investor."

The big Wall Street houses cannot even imagine that such customers exist since they defy everything the houses believe in and hold true. This successful firm has now been widely publicized. It is on every list of large and growing Stock Exchange firms. Yet the senior people in the big firms have not yet accepted that their competitor exists, let alone that it is successful.

Behind the incongruity between actual and perceived reality, there always lies an element of intellectual arrogance, of intellectual rigor and dogmatism. "It is I, not they, who know what poor people can afford," the Japanese industrialist in effect asserted. "People behave according to economic rationality, as every good Marxist knows," as Khrushchev implied. This explains why the incongruity is so easily exploited by innovators: they are left alone and undisturbed.

Of all incongruities, that between perceived and actual reality may be the most common. Producers and suppliers almost always misconceive what it is the customer actually buys. They must assume that what represents "value" to the producer and supplier is equally "value" to the customer. To succeed in doing a job, any job, one has to believe in it and take it seriously. People who make cosmetics must believe in them; otherwise, they turn out shoddy products and soon lose their customers. People who run a hospital must believe in health care as an absolute good, or the quality of medical and patient care will deteriorate fast. And yet, no customer ever perceives himself as buying what the producer or supplier delivers. Their expectations and values are always different.

The reaction of the typical producer and supplier is then to complain that customers are "irrational" or "unwilling to pay for quality." Whenever such a complaint is heard, there is reason to assume that the values and expectations the producer or supplier holds to be real are incongruous with the actual values and expectations of customers and clients. Then there is reason to look for an opportunity for innovation that is highly specific, and carries a good chance of success.

I V

INCONGRUITY WITHIN THE RHYTHM OR
LOGIC OF A PROCESS

Twenty-five years or so ago, during the late 1950s, a pharmaceutical company salesman decided that he wanted to go into business for himself. He therefore looked for an incongruity within a process in medical practice. He found one almost immediately. One of the most common surgical operations is the operation for senile cataract in the eye. Over the years the procedure had become refined, routinized and instrumented to the point where it was conducted with the rhythm of a perfectly rehearsed dance—and with total control. But there was one point in this operation that was out of character and out of rhythm: at one phase the eye surgeon had to cut a ligament, to tie blood vessels and so risk bleeding, which then endangered the eye. This procedure was done successfully in more than 99 percent of all operations; indeed, it was not very difficult. But it greatly bothered the surgeons. It forced them to change their rhythm and induced anxiety in them. Eye surgeons, no matter how often they had done the operation, dreaded this one, quick procedure.

The pharmaceutical company salesman—his name is William Connor—found out without much research that an enzyme had been isolated in the 1890s which almost instantaneously dissolves this particular ligament. Only nobody then, sixty years earlier, had been able to store this enzyme even under refrigeration for more than a few short hours. Preservation techniques have, however, made quite a bit of progress since 1890. And so Connor, within a few months, was able by trial and error to find a preservative that gives the enzyme substantial shelf life without destroying its potency. Within a few years, every eye surgeon in the world was using Connor's patented compound. Twenty years later he sold his company, Alcon Laboratories, to one of the multinationals for a very large amount.

And another telling example:

O. M. Scott & Co. is the leader among American producers of lawn-care products: grass seed, fertilizer, pesticides, and so on. Though it is now a subsidiary of a large corporation (ITT), it attained leadership while a small independent company in fierce competition with firms many times its size, ranging from Sears, Roebuck to Dow Chemicals. Its products are good but so are those of the competition. Its leadership rests on a simple, mechanical gadget called a Spreader, a small, lightweight wheelbarrow with holes that can be set to allow the proper quantities of Scott's products to pass through in an even flow. Products for the lawn all claim to be "scientific" and are compounded on the basis of extensive tests. All prescribe in meticulous detail how much of the stuff should be applied, given soil conditions and temperatures. All try to convey to the consumer that growing a lawn is "precise," "controlled," if not "scientific." But before the Scott Spreader, no supplier of lawn-care products gave the customer a tool to control the process.

And without such a tool, there was an internal incongruity in the logic of the process that upset and frustrated customers.

Does the identification of such internal incongruity within a process rest on "intuition" and on accident? Or can it be organized and systematized?

William Connor is said to have started out by asking surgeons where they felt uncomfortable about their work. O. M. Scott grew from a tiny local seed retailer into a fair-sized national company because it asked dealers and customers what they missed in available products. Then it designed its product line around the Spreader.

The incongruity within a process, its rhythm or its logic, is not a very subtle matter. Users are always aware of it. Every eye surgeon knew about the discomfort he felt when he had to cut eye muscle—and talked about it. Every hardware-store clerk knew about the frustration of his lawn customers—and talked about it. What was lacking, however, was someone willing to listen, somebody who took seriously what everybody proclaims: That the purpose of a product or a service is to satisfy the customer. If this axiom is accepted and acted upon, using incongruity as an opportunity for innovation becomes fairly easy—and highly effective.

There is, however, one serious limitation. The incongruity is usually available only to people within a given industry or service. It is not something that somebody from the outside is likely to spot, to understand, and hence is able to exploit.

5

Source: Process Need

"Opportunity is the source of innovation" has been the leitmotif of the preceding chapters. But an old proverb says, "Necessity is the mother of invention." This chapter looks at *need* as a source of innovation, and indeed as a major innovative opportunity.

The need we shall discuss as a source of innovative opportunity is a very specific one: I call it "process need." It is not vague or general but quite concrete. Like the unexpected, or the incongruities, it exists within the process of a business, an industry, or a service. Some innovations based on process need exploit incongruities, others demographics. Indeed, process need, unlike the other sources of innovation, does not start out with an event in the environment, whether internal or external. It starts out with the job to be done. It is task-focused rather than situation-focused. It perfects a process that already exists, replaces a link that is weak, redesigns an existing old process around newly available knowledge. Sometimes it makes possible a process by supplying the "missing link."

In innovations that are based on process need, everybody in the organization always knows that the need exists. Yet usually no one does anything about it. However, when the innovation appears, it is immediately accepted as "obvious" and soon becomes "standard."

One example has been mentioned earlier in Chapter 4. It is William Connor's conversion of the enzyme that dissolves a ligament in cataract surgery of the eye from a textbook curiosity into an indispensable product. The process of cataract surgery itself was a very old one. The enzyme to perfect the process had been known for decades. The innovation was the preservative to keep the enzyme fresh under refrigeration. Once that process need had been satisfied, no eye surgeon could possibly imagine doing without Connor's compound.

Very few innovations based on process need are so sharply focused

69

as this one, in which formulating the need right away produced the required solution. But in their essentials, most, if not all, innovations based on process need have the same elements.

Here is another example of a similar process-need innovation.

Ottmar Mergenthaler designed the linotype for typesetting in 1885. During the preceding decades, printed materials of all kinds—magazines, newspapers, books—had all been growing at an exponential rate with the spread of literacy and the development of transportation and communication. All the other elements of the printing process had already changed. There were high-speed printing presses, for instance, and paper was being made on high-speed paper machines. Only typesetting had gone unchanged from the days of Gutenberg four hundred years earlier. It remained slow and expensive manual work, requiring high skill and long years of apprenticeship. Mergenthaler, like Connor, defined what was needed: a keyboard that would make possible the mechanical selection of the right letter from the typefont; a mechanism to assemble the letters and to adjust them in a line; and—the most difficult, by the way—a mechanism to return each letter to its proper receptacle for future use. Each of these required several years of hard work and considerable ingenuity. But none required new knowledge, let alone new science. Mergenthaler's linotype became the "standard" in less than five years, despite vigorous resistance from the old craftsmen-typesetters.

In both these cases—William Connor's enzyme and the linotype machine—the process need was based on an incongruity in the process. Demographics, however, are very often an equally powerful source of process need and an opportunity for process innovation.

In 1909 or thereabouts a statistician at the Bell Telephone System projected two curves fifteen years ahead: the curve for American population growth and the curve for the number of people required as central-station operators to handle the growing volume of telephone calls. These projections showed that every American woman between age seventeen and sixty would have to work as a switchboard operator by the year 1925 or 1930 if the manual system of handling calls were to be continued. Two years later, Bell engineers had designed and put into service the first automatic switchboard.

Similarly, the present rush into robotics is largely the result of a process need caused by demographics. Most of the knowledge has been around for years. But until the consequences of the "baby bust" became apparent to major manufacturers in the industrial countries, especially in Japan and the United States, the need to replace semi-skilled assembly-line labor with machines was not felt. The Japanese are not ahead in robotics because of technical superiority; their designs have mostly come from the United States. But the Japanese had their "baby bust" four or five years earlier than America and almost ten years earlier than West Germany. It took the Japanese just as long as it did the Americans or the Germans—ten years—to realize that they were facing a labor shortage. But these ten years started in Japan a good deal sooner than in the United States, and in West Germany the ten years are still not quite over as these lines are being written.

Mergenthaler's linotype was also in large measure the result of demographic pressures. With the demand for printed materials exploding, the supply of typesetters requiring an apprenticeship of six to eight years was fast becoming inadequate, and wages for typesetters were skyrocketing. As a result, printers became conscious of the "weak link" but also willing to pay good money for a machine that replaced five very expensive craftsmen with one semi-skilled machine operator.

Incongruities and demographics may be the most common causes of a process need. But there is another category, far more difficult and risky yet in many cases of even greater importance: what is now being called program research (as contrasted with the traditional "pure research" of scientists). There is a "weak link" and it is definable, indeed, clearly seen and acutely felt. But to satisfy the process need, considerable *new knowledge* has to be produced.

Very few inventions have succeeded faster than photography. Within twenty years after its invention, it had become popular worldwide. Within twenty years or so, there were great photographers in every country; Mathew Brady's photographs of the American Civil War are still unsurpassed. By 1860, every bride had to have her photograph taken. Photography was the first Western technology to invade Japan, well before the Meiji Restoration and at a time when Japan otherwise was still firmly closed to foreigners and foreign ideas.

Amateur photographers were fully established by 1870. But the available technology made things difficult for them. Photography required heavy and fragile glass plates, which had to be lugged around and treated with extreme care. It required an equally heavy camera, long preparations before a picture could be taken, elaborate settings, and so on. Everybody knew this. Indeed, the photography magazines of the time—and photography magazines were among the first specialty mass magazines—are full of complaints about the extreme difficulty of taking photographs and of suggestions what to do. But the problems could not be solved with the science and technology available in 1870.

By the mid-1880s, however, new knowledge had become available which then enabled George Eastman, the founder of Kodak, to replace the heavy glass plates with a cellulose film weighing practically nothing and impervious even to very rough handling, and to design a lightweight camera around his film. Within ten years, Eastman Kodak had taken world leadership in photography, which it still retains.

"Program research" is often needed to convert a process from potential into reality. Again, the need must be felt, and it must be possible to identify what is needed. Then the new knowledge has to be produced. The prototype innovator for this kind of process-need innovation was Edison (see also Chapter 9). For twenty-odd years, everybody had known that there was going to be an "electric power industry." For the last five or six years of that period, it had become abundantly clear what the "missing link" was: the light bulb. Without it, there could be no electric power industry. Edison defined the new knowledge needed to convert this potential electric power industry into an actual one, went to work, and had a light bulb within two years.

Program research to convert a potential into reality has become the central methodology of the first-rate industrial research laboratory and, of course, of research for defense, for agriculture, for medicine, and for environmental protection.

Program research sounds big. To many people it means "putting a man on the moon" or finding a vaccine against polio. But its most successful applications are in small and clearly defined projects—the smaller and the more sharply focused the better. Indeed, the best example—and perhaps the best single example of successful process need–based innovation—is a very small one, the highway reflector that cut the Japanese automobile accident rate by almost two-thirds.

As late as 1965, Japan had almost no paved roads outside of the big cities. But the country was rapidly shifting to the automobile, so the government frantically paved the roads. Now automobiles could—and did—travel at high speed. But the roads were the same old ones that had been laid down by the oxcarts of the tenth century—barely wide enough for two cars to pass, full of blind corners and hidden entrances, and with junctions every few kilometers at which half a dozen roads meet at every conceivable angle. Accidents began to mount at an alarming rate, especially at night. Press, radio and TV, and the opposition parties in Parliament soon began to clamor for the government to "do something." But, of course, rebuilding the roads was out of the question; it would have taken twenty years anyhow. And a massive publicity campaign to make automobilists "drive carefully" had the result such campaigns generally have, namely, none at all.

A young Japanese, Tamon Iwasa, seized on this crisis as an innovative opportunity. He redesigned the traditional highway reflector so that the little glass beads that serve as its mirrors could be adjusted to reflect the headlights of oncoming cars from any direction onto any direction. The government rushed to install Iwasa reflectors by the hundreds of thousands. And the accident rate plummeted.

To take another example.

World War I had created a public in the United States for national and international news. Everybody was aware of this. Indeed, the newspapers and magazines of those early post–World War I years are full of discussions as to how this need could be satisfied. But the local newspaper could not do the job. Several leading publishers tried, among them *The New York Times;* none of them succeeded. Then Henry Luce identified the process need and defined what was required to satisfy it. It could not be a local publication, it had to be a national one, otherwise, there would be neither enough readers nor enough advertisers. And it could not be a daily—there was not enough news of interest to a large public. The development of the editorial format was then practically dictated by these specifications. When *Time* magazine came out as the first news magazine in the world, it was an immediate success.

These examples, and especially the Iwasa story, show that successful innovations based on process needs require five basic criteria:

— A self-contained process;
— One "weak" or "missing" link;
— A clear definition of the objective;
— That the specifications for the solution can be defined clearly;
— Widespread realization that "there ought to be a better way," that is, high receptivity.

There are, however, some important caveats.
1. The need must be *understood*. It is not enough for it to be

"felt." Otherwise one cannot define the specifications for the solution.

We have known, for instance, for several hundred years that mathematics is a problem subject in school. A small minority of students, certainly no more than one-fifth, seem to have no difficulty with mathematics and learn it easily. The rest never really learn it. It is possible, of course, to drill a very much larger percentage to pass mathematics tests. The Japanese do this through heavy emphasis on the subject. But that does not mean that Japanese children learn mathematics. They learn to pass the tests and then immediately forget mathematics. Ten years later, by the time they are in their late twenties, Japanese do just as poorly on mathematics tests as do westerners. In every generation there is a mathematics teacher of genius who somehow can make even the untalented learn, or at least learn a good deal better. But nobody has ever been able, then, to replicate what this one person does. The need is acutely felt, but we do not understand the problem. Is it a lack of native ability? Is it that we are using the wrong methods? Are there psychological and emotional problems? No one knows the answer. And without understanding the problem, we have not been able to find any solution.

2. We may even understand a process and still not have the knowledge to do the job. The preceding chapter told of the clear and understood incongruity in paper making: to find a process that is less wasteful and less uneconomical than the existing one. For a century, able people have worked on the problem. We know exactly what is needed: polymerization of the lignin molecule. It should be easy—we have polymerized many molecules that are similar. But we lack the knowledge to do it, despite a hundred years of assiduous work by well-trained people. One can only say, "Let's try something else."

3. The solution must fit the way people do the work and want to do it. Amateur photographers had no psychological investment in the complicated technology of the early photographic process. All they wanted was to get a decent photograph, as easily as possible. They were receptive, therefore, to a process that took the labor and skill out of taking pictures. Similarly, eye surgeons were interested only in an elegant, logical, bloodless process. An enzyme that gave this to them therefore satisfied their expectations and values.

But here is an example of an innovation based on a clear and substantial process need that apparently does not quite fit, and therefore has not been readily accepted.

For many years the information required by a number of professionals such as lawyers, accountants, engineers, and physicians has grown much faster than the capacity to find it. Professionals have been complaining that they have to spend more and more time hunting for information in the law library, in handbooks and textbooks, in looseleaf services, and so on. One would therefore expect a "databank" to be an immediate success. It gives the professionals immediate information through a computer program and a display terminal: court decisions for the lawyers, tax rulings for the accountants, information on drugs and poisons for the physicians. Yet these services have found it very hard to gather enough subscribers to break even. In some cases, such as Lexis, a service for lawyers, it has taken more than ten years and huge sums of money to get subscribers. The reason is probably that the databanks make it *too* easy. Professionals pride themselves on their "memory," that is, on their ability either to remember the information they need or to know where to find it. "You have to remember the court decisions you need and where to find them," is still the injunction the beginning lawyer gets from the seniors. So the databank, however helpful in the work and however much time and money it saves, goes against the very values of the professional. "What would you need *me* for if it can be looked up?" an eminent physician once said when asked by one of his patients why he did not use the service that would give him the information to check and confirm his diagnosis, and then decide which alternative method of treatment might be the best in a given case.

Opportunities for innovation based on process need can be found systematically. This is what Edison did for electricity and electronics. This is what Henry Luce did while still an undergraduate at Yale. This is what William Connor did. In fact, the area lends itself to systematic search and analysis.

But once a process need has been found, it has to be tested against the five basic criteria given above. Then, finally, the process need opportunity has to be tested also against the three constraints. Do we understand what is needed? Is the knowledge available or can it be procured within the "state of the art"? And does the solution fit, or does it violate the mores and values of the intended users?

6

Source: Industry and Market Structures

Industry and market structures sometimes last for many, many years and seem completely stable. The world aluminum industry, for instance, after one century is still led by the Pittsburgh-based Aluminum Company of America which held the original patents, and by its Canadian offspring, Alcan of Montreal. There has only been one major newcomer in the world's cigarette industry since the 1920s, the South African Rembrandt group. And in an entire century only two newcomers have emerged as leading electrical apparatus manufacturers in the world: Philips in Holland and Hitachi in Japan. Similarly no major new retail chain emerged in the United States for forty years, between the early twenties when Sears, Roebuck began to move from mail order into retail stores, and the mid-sixties when an old dime-store chain, Kresge, launched the K-Mart discount stores. Indeed, industry and market structures appear so solid that the people in an industry are likely to consider them foreordained, part of the order of nature, and certain to endure forever.

Actually, market and industry structures are quite brittle. One small scratch and they disintegrate, often fast. When this happens, every member of the industry has to act. To continue to do business as before is almost a guarantee of disaster and might well condemn a company to extinction. At the very least the company will lose its leadership position; and once lost, such leadership is almost never regained. But a change in market or industry structure is also a major opportunity for innovation.

In industry structure, a change requires entrepreneurship from every member of the industry. It requires that each one ask anew: "What is our business?" And each of the members will have to give a different, but above all a new, answer to that question.

I

THE AUTOMOBILE STORY

The automobile industry in the early years of this century grew so fast that its markets changed drastically. There were four different responses to this change, all of them successful. The early industry through 1900 had basically been a provider of a luxury product for the very rich. By then, however, it was outgrowing this narrow market with a rate of growth that doubled the industry's sales volume every three years. Yet the existing companies all still concentrated on the "carriage trade."

One response to this was the British company, Rolls-Royce, founded in 1904. The founders realized that automobiles were growing so plentiful as to become "common," and set out to build and sell an automobile which, as an early Rolls-Royce prospectus put it, would have "the cachet of royalty." They deliberately went back to earlier, already obsolete, manufacturing methods in which each car was machined by a skilled mechanic and assembled individually with hand tools. And then they promised that the car would never wear out. They designed it to be driven by a professional chauffeur trained by Rolls-Royce for the job. They restricted sales to customers of whom they approved—preferably titled ones, of course. And to make sure that no "riff-raff" bought their car, they priced the Rolls-Royce as high as a small yacht, at about forty times the annual income of a skilled mechanic or prosperous trades-man.

A few years later in Detroit, the young Henry Ford also saw that the market structure was changing and that automobiles in America were no longer a rich man's toy. His response was to design a car that could be totally mass-produced, largely by semi-skilled labor, and that could be driven by the owner and repaired by him. Contrary to legend, the 1908 Model T was not "cheap": it was priced at a little over what the world's highest-priced skilled mechanic, the American one, earned in a full year. (These days, the cheapest new car on the American market costs about one-tenth of what an unskilled assembly-line worker gets in wages and benefits in a year.) But the Model T cost one-fifth of the cheapest model then on the market and was infinitely easier to drive and to maintain.

Another American, William Crapo Durant, saw the change in market structure as an opportunity to put together a professionally managed large automobile company that would satisfy all segments of what he foresaw would be a huge "universal" market. He founded General Motors in 1905, began to buy existing automobile companies, and integrated them into a large modern business.

A little earlier, in 1899, the young Italian Giovanni Agnelli had seen that the automobile would become a military necessity, especially as a staff car for officers. He founded FIAT in Turin, which within a few years became the leading supplier of staff cars to the Italian, Russian, and Austro-Hungarian armies.

Market structures in the world automobile industry changed once again between 1960 and 1980. For forty years after World War I, the automobile industry had consisted of national suppliers dominating national markets. All one saw on Italy's roads and parking lots were Fiats and a few Alfa Romeos and Lancias; outside of Italy, these makes were fairly rare. In France, there were Renaults, Peugeots, and Citroens; in Germany, Mercedes, Opels, and the German Fords; in the United States, GM cars, Fords, and Chryslers. Then around 1960 the automobile industry all of a sudden became a "global" industry.

Different companies reacted quite differently. The Japanese, who had remained the most insular and had barely exported their cars, decided to become world exporters. Their first attempt at the U.S. market in the late sixties was a fiasco. They regrouped, thought through again what their policy should be, and redefined it as offering an American-type car with American styling, American comfort, and American performance characteristics, but smaller, with better fuel consumption, much more rigorous quality control and, above all, better customer service. And when they got a second chance with the petroleum panic of 1979, they succeeded brilliantly. The Ford Motor Company, too, decided to go "global" through a "European" strategy. Ten years later, in the mid-seventies, Ford had become a strong contender for the number one spot in Europe.

Fiat decided to become a European rather than merely an Italian company, aiming to be a strong number two in every important European country while retaining its primary position in Italy. General Motors at first decided to remain American and to retain its traditional 50 percent share of the American market, but in such a way as to reap something like 70 percent of all profits from automobile sales in North

America. And it succeeded. Ten years later, in the mid-seventies, GM shifted gears and decided to contend with Ford and Fiat for leadership in Europe—and again it succeeded. In 1983–84, GM, it would seem, decided finally to become a truly global company and to link up with a number of Japanese; first with two smaller companies, and in the end with Toyota. And Mercedes in West Germany decided on yet another strategy—again a global one—where it limited itself to narrow segments of the world market, to luxury cars, taxicabs, and buses.

All these strategies worked reasonably well. Indeed, it is impossible to say which one worked better than another. But the companies that refused to make hard choices, or refused to admit that anything much was happening, fared badly. If they survive, it is only because their respective governments will not let them go under.

One example is, of course, Chrysler. The people at Chrysler knew what was happening—everybody in the industry did. But they ducked instead of deciding. Chrysler might have chosen an "American" strategy and put all its resources into strengthening its position within the United States, still the world's largest automobile market. Or it might have merged with a strong European firm and aimed at taking third place in the world's most important automobile markets, the United States and Europe. It is known that Mercedes was seriously interested —but Chrysler was not. Instead, Chrysler frittered away its resources on make-believe. It acquired defeated "also-rans" in Europe to make itself look multinational. But this, while giving Chrysler no additional strength, drained its resources and left no money for the investment needed to give Chrysler a chance in the American market. When the day of reckoning came after the petroleum shock of 1979, Chrysler had nothing in Europe and not much more in the United States. Only the U.S. government saved it.

The story is not much different for British Leyland, once Britain's largest automobile company and a strong contender for leadership in Europe; nor for the big French automobile company, Peugeot. Both refused to face up to the fact that a decision was needed. As a result, they rapidly lost both market position and profitability. Today all three —Chrysler, British Leyland, and Peugeot—have become more or less marginal.

But the most interesting and important examples are those of much smaller companies. Every one of the world's automobile manufacturers, large or small, has had to act or face permanent eclipse. However, three small and quite marginal companies saw in this a major opportunity to innovate: Volvo, BMW, and Porsche.

Around 1960, when the automobile industry market suddenly changed, the informed betting was heavily on the disappearance of these three companies during the coming "shakeout." Instead, all three have done well and have created for themselves market niches in which they are the leaders. They have done so through an innovative strategy which, in effect, has reshaped them into different businesses. Volvo in 1965 was small, struggling and barely breaking even. For a few critical years, it did lose large amounts of money. But Volvo went to work reinventing itself, so to speak. It became an aggressive worldwide marketer—especially strong in the United States—of what one might call the "sensible" car; not very luxurious, far from low-priced, not at all fashionable, but sturdy and radiating common sense and "better value." Volvo has marketed itself as the car for professionals who do not need to demonstrate how successful they are through the car they drive, but who value being known for their "good judgment."

BMW, equally marginal in 1960 if not more so, has been equally successful, especially in countries like Italy and France. It has marketed itself as the car for "young comers," for people who want to be taken as young but who already have attained substantial success in their work and profession, people who want to demonstrate that they "know the difference" and are willing to pay for it. BMW is unashamedly a luxury car for the well-to-do, but it appeals to those among the affluent who want to appear "nonestablishment." Whereas Mercedes and Cadillac are the cars for company presidents and for heads of state, BMW is *muy macho,* and bills itself as the "ultimate driving machine."

Finally Porsche (originally a Volkswagen with extra styling) repositioned itself as *the* sports car, the one and only car for those who still do not want transportation but excitement in an automobile.

But those smaller automobile manufacturers who did not innovate and present themselves differently in what is, in effect, a different business—those who continued their established ways—have become casualties. The British MG, for instance, was thirty years ago what Porsche has now become, the sports car *par excellence.* It is almost extinct by now. And where is Citroen? Thirty years ago it was the car that had the solid innovative engineering, the sturdy construction, the middle-class reliability. Citroen would have seemed to be ideally positioned for the market niche Volvo has taken over. But Citroen failed to think through its business and to innovate; as a result, it has neither product nor strategy.

II

THE OPPORTUNITY

A change in industry structure offers exceptional opportunities, highly visible and quite predictable to outsiders. But the insiders perceive these same changes primarily as threats. The outsiders who innovate can thus become a major factor in an important industry or area quite fast, and at relatively low risk.

Here are some examples.

In the late 1950s three young men met, almost by accident, in New York City. Each of them worked for financial institutions, mostly Wall Street houses. They found themselves in agreement on one point: the securities business—unchanged since the Depression twenty years earlier—was poised for rapid structural change. They decided that this change had to offer opportunities. So they systematically studied the financial industry and the financial markets to find an opportunity for newcomers with limited capital resources and practically no connections. The result was a new firm: Donaldson, Lufkin & Jenrette. Five years after it had been started in 1959, it had become a major force on Wall Street.

What these three young men found was that a whole new group of customers was emerging fast: the pension fund administrators. These new customers did not need anything that was particularly difficult to supply, but they needed something different. And no existing firm had organized itself to give it to them. Donaldson, Lufkin & Jenrette established a brokerage firm to focus on these new customers and to give them the "research" they needed.

About the same time, another young man in the securities business also realized that the industry was in the throes of structural change and that this could offer him an opportunity to build a different securities business of his own. The opportunity he found was "the intelligent investor" mentioned earlier. On this, he then built what is now a big and still fast-growing firm.

During the early or mid-sixties, the structure of American health care began to change very fast. Three young people, the oldest not quite thirty, then working as junior managers in a large Midwestern hospital, decided that this offered them an opportunity to start their own innovative business. They concluded that hospitals would increasingly need expertise in running such housekeeping services as kitchen, laundry, maintenance, and so on. They systematized the work to be done. Then they offered contracts to hospitals under which their new firm would put in its own trained people to run these services, with the fee a portion of the resultant savings. Twenty years later, this company billed almost a billion dollars of services.

The final case is that of the discounters like MCI and Sprint in the American long-distance telephone market. They were total outsiders; Sprint, for instance, was started by a railroad, the Southern Pacific. These outsiders began to look for the chink in Bell System's armor. They found it in the pricing structure of long-distance services. Until World War II, long-distance calls had been a luxury confined to government and large businesses, or to emergencies such as a death in the family. After World War II, they became commonplace. Indeed, they became the growth sector of telecommunications. But under pressure from the regulatory authorities for the various states which control telephone rates, the Bell System continued to price long-distance as a luxury, way above costs, with the profits being used to subsidize local service. To sweeten the pill, however, the Bell System gave substantial discounts to large buyers of long-distance service.

By 1970, revenues from long-distance service had come to equal those from local service and were fast outgrowing them. Still, the original price structure was maintained. And this is what the newcomers exploited. They signed up for volume service at the discount and then retailed it to smaller users, splitting the discount with them. This gave them a substantial profit while also giving their subscribers long-distance service at substantially lower cost. Ten years later, in the early eighties, the long-distance discounters handled a larger volume of calls than the entire Bell System had handled when the discounters first started.

These cases would just be anecdotes except for one fact: each of the innovators concerned *knew* that there was a major innovative opportunity in the industry. Each was reasonably sure that an innovation would succeed, and succeed with minimal risk. How could they be so sure?

III

WHEN INDUSTRY STRUCTURE CHANGES

Four near-certain, highly visible indicators of impending change in industry structure can be pinpointed.

1. The most reliable and the most easily spotted of these indicators is rapid growth of an industry. This is, in effect, what each of the above examples (but also the automobile industry examples) have in common. If an industry grows significantly faster than economy or population, it can be predicted with high probability that its structure will change drastically—at the very latest by the time it has doubled in volume. Existing practices are still highly successful, so nobody is inclined to tamper with them. Yet they are becoming obsolete. Neither the people at Citroen nor those at Bell Telephone were willing to accept this, however—which explains why "newcomers," "outsiders," or former "second-raters" could beat them in their own markets.

2. By the time an industry growing rapidly has doubled in volume, the way it perceives and services its market is likely to have become inappropriate. In particular, the ways in which the traditional leaders define and segment the market no longer reflect reality, they reflect history. Yet reports and figures still represent the traditional view of the market. This is the explanation for the success of two such different innovators as Donaldson, Lufkin & Jenrette and the Midwestern "intelligent investor" brokerage house. Each found a segment that the existing financial services institutions had not perceived and therefore did not serve adequately; the pension funds because they were too new, the "intelligent investor" because he did not fit the Wall Street stereotype.

But the hospital management story is also one of traditional aggregates no longer being adequate after a period of rapid growth. What grew in the years after World War II were the "paramedics," that is, the hospital professions: X-Ray, pathology, the medical lab, therapists of all kinds, and so on. Before World War II these had barely existed. And hospital administration itself became a profession. The traditional "housekeeping" services, which had dominated hospital operations in earlier times, thus steadily became a problem for the administrator, proving increasingly difficult and costly as hospital employees, especially the low-paid ones, began to unionize.

And the case of the book chains reported earlier (in Chapter 3) is also a story of structural change because of rapid growth. What neither the publishers nor the traditional American bookstores realized was that new customers, the "shoppers," were emerging side by side with the old customers, the traditional readers. The traditional bookstore simply did not perceive these new customers and never attempted to serve them.

But there is also the tendency if an industry grows very fast to become complacent and, above all, to try to "skim the cream." This is what the Bell System did with respect to long-distance calls. The sole result is to invite competition (on this see also Chapter 17).

Yet another example is to be found in the American art field. Before World War II, museums were considered "upper-class." After World War II, going to museums became a middle-class habit; in city after city new museums were founded. Before World War II, collecting art was something a few very rich people did. After World War II, collecting all kinds of art became increasingly popular, with thousands of people getting into the act, some of them people of fairly limited means.

One young man working in a museum saw this as an opportunity for innovation. He found it in the most unexpected place—in fact, in a place he had never heard of before, insurance. He established himself as an insurance broker specializing in art and insuring both museums and collectors. Because of his art expertise, the underwriters in the major insurance companies, who had been reluctant to insure art collections, became willing to take the risk, and at premiums up to 70 percent below those charged before. This young man now has a large insurance brokerage firm.

3. Another development that will predictably lead to sudden changes in industry structure is the convergence of technologies that hitherto were seen as distinctly separate.

One example is that of the private branch exchange (PBX), that is, the switchboard for offices and other large telephone users. Basically, all the scientific and technical work on this in the United States has been done by Bell Labs, the research arm of the Bell System. But the main beneficiaries have been a few newcomers such as ROLM Corporation. In the new PBX, two different technologies converge: telephone technology and computer technology. The PBX can be seen as a telecommunications instrument that uses a computer, or as a computer that is being used in telecommunications. Technically, the Bell System would have been perfectly capable of handling this—in fact, it has all along been a computer pioneer. In its view of the market, however, and of the user, Bell System saw the computer as something totally different and far away. While it designed and actually introduced a computer-type PBX, it never pushed it. As a result, a total newcomer has become a major competitor. In fact, ROLM, started by four young engineers, was founded to build a small computer for fighter aircraft, and only stumbled by accident into the telephone business. The Bell System now has not much more than one-third of that market, despite its technical leadership.

4. An industry is ripe for basic structural change if the way in which it does business is changing rapidly.

Thirty years ago, the overwhelming majority of American physicians practiced on their own. By 1980, only 60 percent were doing so. Now, 40 percent (and 75 percent of the younger ones) practice in a group, either in a partnership or as employees of a Health Maintenance Organization or a hospital. A few people who saw what was happening early on, around 1970, realized that it offered an opportunity for innovation. A service company could design the group's office, tell the physicians what equipment they needed, and either manage their group practice for them or train their managers.

Innovations that exploit changes in industry structure are particularly effective if the industry and its markets are dominated by one very large manufacturer or supplier, or by a very few. Even if there is no true monopoly, these large, dominant producers and suppliers, having been successful and unchallenged for many years, tend to be arrogant. At first they dismiss the newcomer as insignificant and, indeed, amateurish. But even when the newcomer takes a larger and larger share of their business, they find it hard to mobilize themselves for counteraction. It took the Bell System almost ten years before it first responded to the long-distance discounters and to the challenge from the PBX manufacturers.

Equally sluggish, however, was the response of the American producers of aspirin when the "non-aspirin aspirins"—Tylenol and Datril —first appeared (on this see also Chapter 17). Again, the innovators diagnosed an opportunity because of an impending change in industry structure, based very largely on rapid growth. There was no reason whatever why the existing aspirin manufacturers, a very small number of very large companies, could not have brought out "non-aspirin aspirin" and sold it effectively. After all, the dangers and limitations of aspirin were no secret; medical literature was full of them. Yet, for the first five or eight years, the newcomers had the market to themselves.

Similarly, the United States Postal Service did not react for many years to innovators who took away larger and larger chunks of the most profitable services. First, United Parcel Service took away ordinary parcel post; then Emery Air Freight and Federal Express took away the even more profitable delivery of urgent or high-value merchandise and letters. What made the Postal Service so vulnerable was its rapid growth. Volume grew so fast that it neglected what seemed to be minor categories, and thus practically delivered an invitation to the innovators.

Again and again when market or industry structure changes, the producers or suppliers who are today's industry leaders will be found neglecting the fastest-growing market segments. They will cling to practices that are rapidly becoming dysfunctional and obsolete. The new growth opportunities rarely fit the way the industry has "always" approached the market, been organized for it, and defines it. The innovator in this area therefore has a good chance of being left alone. For some time, the old businesses or services in the field will still be doing well serving the old market the old way. They are likely to pay little attention to the new challenge, either treating it with condescension or ignoring it altogether.

But there is one important caveat. It is absolutely essential to keep the innovation in this area simple. Complicated innovations do not work. Here is one example, the most intelligent business strategy I know of and one of the most dismal failures.

Volkswagen triggered the change which converted the automobile industry around 1960 into a global market. The Volkswagen Beetle was the first car since the Model T forty years earlier that became a truly international car. It was as ubiquitous in the United States as it was in its native Germany, and as familiar in Tanganyika as it was in the Solomon Islands. And yet Volkswagen missed the opportunity it had created itself—primarily by being *too* clever.

By 1970, ten years after its breakthrough into the world market, the Beetle was becoming obsolete in Europe. In the United States, the Beetle's second-best market, it still sold moderately well. And in Brazil, the Beetle's third-largest market, it apparently still had substantial growth ahead. Obviously, new strategy was called for.

The chief executive officer of Volkswagen proposed switching the German plants entirely to the new model, the successor to the Beetle, which the German plants would also supply to the United States market. But the continuing demand for Beetles in the United States would be satisfied out of Brazil, which would then give Volkswagen do Brasil the needed capacity to enlarge its plants and to maintain for another ten years the Beetle's leadership in the growing Brazilian market. To assure the American customers of the "German quality" that was one of the Beetle's main attractions, the critical parts such as engines and transmissions for all cars sold in North America would, however, still be made in Germany, with the finished car for the North American market then assembled in the United States.

In its way, this was the first genuinely global strategy, with different parts to be made in different countries and assembled in different places according to the needs of different markets. Had it worked, it would have been the right strategy, and a highly innovative one at that. It was killed primarily by the German labor unions. "Assembling Beetles in the United States means exporting German jobs," they said, "and we won't stand for it." But the American dealers were also doubtful about a car that was "made in Brazil," even though the critical parts would still be "made in Germany." And so Volkswagen had to give up its brilliant plan.

The result has been the loss of Volkswagen's second market, the United States. Volkswagen, and not the Japanese, should have had the small car market when small cars became all the rage after the fall of the Shah of Iran triggered the second petroleum panic. Only the Germans had no product. And when, a few years later, Brazil went into a severe economic crisis and automobile sales dropped, Volkswagen do Brasil got into difficulties. There were no export customers for the capacity it had had to build there during the seventies.

The specific reasons why Volkswagen's brilliant strategy failed—to the point where the long-term future of the company may have become problematical—are secondary. The moral of the story is that a "clever" innovative strategy always fails, particularly if it is aimed at exploiting an opportunity created by a change in industry structure. Then only the very simple, specific strategy has a chance of succeeding.

7

Source: Demographics

The unexpected; incongruities; changes in market and industry structure; and process needs—the sources of innovative opportunity discussed so far in Chapters 3 through 6—manifest themselves within a business, an industry, or a market. They may actually be symptoms of changes outside, in the economy, in society, and in knowledge. But they show up internally.

The remaining sources of innovative opportunity:

— Demographics;
— Changes in perception, meaning, and mood;
— New knowledge

are external. They are changes in the social, philosophical, political, and intellectual environment.

I

Of all external changes, demographics—defined as changes in population, its size, age structure, composition, employment, educational status, and income—are the clearest. They are unambiguous. They have the most predictable consequences.

They also have known and almost certain lead times. Anyone in the American labor force in the year 2000 is alive by now (though not necessarily living in the United States; a good many of America's workers fifteen years hence may now be children in a Mexican *pueblo,* for example). All people reaching retirement age in 2030 in the developed countries are already in the labor force, and in most cases in the occupational group in which they will stay until they retire or die. And the educational attainment of the people now in their early or mid-twenties will largely determine their career paths for another forty years.

Demographics have major impact on what will be bought, by whom, and in what quantities. American teenagers, for instance, buy a good many pairs of cheap shoes a year; they buy for fashion, not durability, and their purses are limited. The same people, ten years later, will buy very few pairs of shoes a year—a sixth as many as they bought when they were seventeen—but they will buy them for comfort and durability first and for fashion second. People in their sixties and seventies in the developed countries—that is, people in their early retirement years —form the prime travel and vacation market. Ten years later the same people are customers for retirement communities, nursing homes, and extended (and expensive) medical care. Two-earner families have more money than they have time, and spend accordingly. People who have acquired extensive schooling in their younger years, especially professional or technical schooling, will, ten to twenty years later, become customers for advanced professional training.

But people with extensive schooling are also available primarily for employment as knowledge workers. Even without competition from low-wage countries with tremendous surpluses of young people trained only for unskilled or semi-skilled manual jobs—the surge of young people in the Third World countries resulting from the drop in infant mortality after 1955—the industrially developed countries of the West and of Japan would have had to automate. Demographics alone, the combined effects of the sharp drop in birth rates and of the "educational explosion"—makes it near-certain that traditional manual blue-collar employment in manufacturing in developed countries, by the year 2010, cannot be more than one-third or less than what it was in 1970. (Though manufacturing production, as a result of automation, may be three to four times what it was then.)

All this is so obvious that no one, one should think, needs to be reminded of the importance of demographics. And indeed businessmen, economists, and politicians have always acknowledged the critical importance of population trends, movements, and dynamics. But they also believed that they did not have to pay attention to demographics in their day-to-day decisions. Population changes—whether in birth rates or mortality rates, in educational attainment, in labor force composition and participation, or in the location and movement of people —were thought to occur so slowly and over such long time spans as to be of little practical concern. Great demographic catastrophes such as the Black Death in Europe in the fourteenth century were admitted to have immediate impacts on society and economy. But otherwise, demographic changes were "secular" changes, of interest to the historian and the statistician rather than to the businessman or the administrator.

This was always a dangerous error. The massive nineteenth-century migration from Europe to the Americas, both North and South, and to Australia and New Zealand, changed the economic and political geography of the world beyond recognition. It created an abundance of entrepreneurial opportunities. It made obsolete the geopolitical concepts on which European politics and military strategies had been based for several centuries. Yet it took place in a mere fifty years, from the mid-1860s to 1914. Whoever disregarded it was likely to be left behind, and fast.

Until 1860, for instance, the House of Rothschild was the world's dominant financial power. The Rothschilds failed, however, to recognize the meaning of the transatlantic migration; only "riff-raff," they thought, would leave Europe. As a result, the Rothschilds ceased to be important around 1870. They had become merely rich individuals. It was J. P. Morgan who took over. His "secret" was to spot the transatlantic migration at its very onset, to understand immediately its significance, and to exploit it as an opportunity by establishing a worldwide bank in New York rather than in Europe, and as the medium for financing the American industries that immigrant labor was making possible. It also took only thirty years, from 1830 to 1860, to transform both western Europe and the eastern United States from rural and farm-based societies into industry-dominated big-city civilizations.

Demographic changes tended to be just as fast, just as abrupt, and to have fully as much impact, in earlier times. The belief that populations changed slowly in times past is pure myth. Or rather, static populations staying in one place for long periods of time have been the exception historically rather than the rule.*

In the twentieth century it is sheer folly to disregard demographics. The basic assumption for our time must be that populations are inherently unstable and subject to sudden sharp changes—and that they are the first environmental factor that a decision maker, whether businessman or politician, analyzes and thinks through. Few issues in this century, for instance, will be as critical to both domestic and international politics as the aging of the population in the developed countries on the

*Here the work of the modern French historians of civilization is definitive.

one hand and the tidal wave of young adults in the Third World on the other hand. Whatever the reasons, twentieth-century societies, both developed and developing ones, have become prone to extremely rapid and radical demographic changes, which occur without advance warning.

The most prominent American population experts called together by Franklin D. Roosevelt predicted unanimously in 1938 that the U.S. population would peak at around 140 million people in 1943 or 1944, and then slowly decline. The American population—with a minimum of immigration—now stands at 240 million. For in 1949, without the slightest advance warning, the United States kicked off a "baby boom" that for twelve years produced unprecedentedly large families, only to turn just as suddenly in 1960 into a "baby bust," producing equally unprecedented small families. The demographers of 1938 were not incompetents or fools; there was just no indication then of a "baby boom."

Twenty years later another American President, John F. Kennedy, called together a group of eminent experts to work out his Latin-American aid and development program, the "Alliance for Progress." Not one of the experts paid attention in 1961 to the precipitous drop in infant mortality which, within another fifteen years, totally changed Latin America's society and economy. The experts also all assumed, without reservation, a rural Latin America. They, too, were neither incompetents nor fools. But the drop in infant mortality in Latin America and the urbanization of society had barely begun at the time.

In 1972 and 1973, the most experienced labor force analysts in the United States still accepted without question that the participation of women would continue to decline as it had done for many years. When the "baby boomers" came on the labor market in record numbers, they worried (quite unnecessarily, as it turned out) where all the jobs for the young males would be coming from. No one asked where jobs would come from for young females—they were not supposed to need any. Ten years later the labor force participation of American women under fifty stood at 64 per cent, the highest rate ever. And there is little difference in labor force participation in this group between married and unmarried women, or between women with and without children.

These shifts are not only dazzlingly sudden. They are often mysterious and defy explanation. The drop in infant mortality in the Third World can be explained in retrospect. It was caused by a convergence of old technologies: the public-health nurse; placing the latrine below the well; vaccination; the wire screen outside the window; and, of very new technologies, antibiotics and pesticides such as DDT. Yet it was totally unpredictable. And what explains the "baby boom" or the "baby bust"? What explains the sudden rush of American women (and of European women as well, though with a lag of a few years) into the labor force? And what explains the rush into the slums of Latin-American cities?

Demographic shifts in this century may be inherently unpredictable, yet they do have long lead times before impact, and lead times, moreover, which are predictable. It will be five years before newborn babies become kindergarten pupils and need classrooms, playgrounds, and teachers. It will be fifteen years before they become important as customers, and nineteen to twenty years before they join the labor force as adults. Populations in Latin America began to grow quite rapidly as soon as infant mortality began to drop. Still the babies who did not die did not become schoolchildren for five or six years, nor adolescents looking for work for fifteen or sixteen years. And it takes at least ten years—usually fifteen—before any change in educational attainments translates itself into labor force composition and available skills.

What makes demographics such a rewarding opportunity for the entrepreneur is precisely its neglect by decision makers, whether businessmen, public-service staffs, or governmental policymakers. They still cling to the assumption that demographics do not change—or do not change fast. Indeed, they reject even the plainest evidence of demographic changes. Here are some fairly typical examples.

By 1970, it had become crystal clear that the number of children in America's schools was going to be 25 to 30 percent lower than it had been in the 1960s, for ten or fifteen years at least. After all, children entering kindergarten in 1970 have to be alive no later than 1965, and the "baby bust" was well established beyond possibility of rapid reversal by that year. Yet the schools of education in American universities flatly refused to accept this. They considered it a law of nature, it seems, that the number of children of school age must go up year after year. And so they stepped up their efforts to recruit students, causing substantial unemployment for graduates a few years later, severe pressure on teachers' salaries, and massive closings of schools of education.

And here are two examples from my own experience. In 1957, I published a forecast that there would be ten to twelve million college students in the United States twenty-five years later, that is, by the mid-seventies. The figure was derived simply by putting together two demographic events that had already happened: the increase in the number of births and the increase in the percentage of young adults going to college. The forecast was absolutely correct. Yet practically every established university pooh-poohed it. Twenty years later, in 1976, I looked at the age figures and predicted that retirement age in the United States would have to be raised to seventy or eliminated altogether within ten years. The change came even faster: compulsory retirement at any age was abolished in California a year later, in 1977, and retirement before seventy for the rest of the country two years later, in 1978. The demographic figures that made this prediction practically certain were well known and published. Yet most so-called experts—government economists, labor-union economists, business economists, statisticians—dismissed the forecast as utterly absurd. "It will never happen" was the all but unanimous response. The labor unions actually proposed at the time lowering the mandatory retirement age to sixty or below.

This unwillingness, or inability, of the experts to accept demographic realities which do not conform to what they take for granted gives the entrepreneur his opportunity. The lead times are known. The events themselves have already happened. But no one accepts them as reality, let alone as opportunity. Those who defy the conventional wisdom and accept the facts—indeed, those who go actively looking for them—can therefore expect to be left alone for quite a long time. The competitors will accept demographic reality, as a rule, only when it is already about to be replaced by a new demographic change and a new demographic reality.

II

Here are some examples of successful exploitation of demographic changes.

Most of the large American universities dismissed my forecast of 10 to 12 million college students by the 1970s as preposterous. But the entrepreneurial universities took it seriously: Pace University, in New York, was one, and Golden Gate University in San Francisco another. They were just as incredulous at first, but they checked the forecast and found that it was valid, and in fact the only rational prediction. They then organized themselves for the additional student enrollment; the traditional, and especially the "prestige" universities, on the other hand, did nothing. As a result, twenty years later these brash newcomers had the students, and when enrollments decreased nationwide as a result of the "baby bust," they still kept on growing.

One American retailer who accepted the "baby boom" was then a small and undistinguished shoe chain, Melville. In the early 1960s just before the first cohorts of the "baby boom" reached adolescence, Melville directed itself to this new market. It created new and different stores specifically for teenagers. It redesigned its merchandise. It advertised and promoted to the sixteen- and seventeen-year-olds. And it went beyond footwear into clothing for teenagers, both female and male. As a result, Melville became one of the fastest-growing and most profitable retailers in America. Ten years later other retailers caught on and began to cater to teenagers—just as the center of demographic gravity started to shift away from them and toward "young adults," twenty to twenty-five years old. By then Melville was already shifting its own focus to that new dominant age cohort.

The scholars on Latin America whom President Kennedy brought together to advise him on the Alliance for Progress in 1961 did not see Latin America's urbanization. But one business, the American retail chain Sears, Roebuck, had seen it several years earlier—not by poring over statistics but by going out and looking at customers in Mexico City and Lima, São Paulo and Bogotá. As a result, Sears in the mid-fifties began to build American-type department stores in major Latin-American cities, designed for a new urban middle class which, while not "rich," was part of the money economy and had middle-class aspirations. Sears became the leading retailer in Latin America within a few years.

And here are two examples of exploiting demographics to innovate in building a highly productive labor force. The expansion of New York's Citibank is largely based on its early realization of the movement of young, highly educated and highly ambitious women into the work force. Most large American employers considered these women a "problem" as late as 1980; many still do. Citibank, almost alone among large employers, saw in them an opportunity. It aggressively recruited them during the 1970s, trained them, and sent them out all over the country as lending officers. These ambitious young women very largely made Citibank into the nation's leading, and its first truly "national" bank. At the same time, a few savings and loan associations (not an industry noted for innovation or venturing) realized that older married women who had earlier dropped out of the labor force when their children were small make high-grade employees when brought back as permanent part-time workers. "Everybody knew" that part-timers are "temporary," and that women who have once left the labor force never come back into it; both were perfectly sensible rules in earlier times. But demographics made them obsolete. The willingness to accept this fact—and again such willingness stemmed not from reading statistics but from going out and looking—has given the savings and loan associations an exceptionally loyal, exceptionally productive work force, particularly in California.

The success of Club Méditerranée in the travel and resort business is squarely the result of exploiting demographic changes: the emergence of large numbers of young adults in Europe and the United States who are affluent and educated but only one generation away from working-class origins. Still quite unsure of themselves, still not self-confident as tourists, they are eager to have somebody with the know-how to organize their vacations, their travel, their fun—and yet they are not really comfortable either with their working-class parents or with older, middle-class people. Thus, they are ready-made customers for a new and "exotic" version of the old teenage hangout.

III

Analysis of demographic changes begins with population figures. But absolute population is the least significant number. Age distribution is far more important, for instance. In the 1960s, it was the rapid increase in the number of young people in most non-Communist developed countries that proved significant (the one notable exception was Great Britain, where the "baby boom" was short-lived). In the 1980s and even more in the 1990s, it will be the drop in the number of young people, the steady increase in the number of early middle-age people (up to forty) and the very rapid increase in the number of old people (seventy and over). What opportunities do these developments offer? What are the values and the expectations, the needs and wants of these various age groups?

The number of traditional college students cannot increase. The most one can hope for is that it will not fall, that the percentage of eighteen- and nineteen-year-olds who stay in school beyond secondary education will increase sufficiently to offset the decline in the total number. But with the increase in the number of people in their mid-thirties and forties who have received a college degree earlier, there are going to be large numbers of highly schooled people who want advanced professional training and retraining, whether as doctors, lawyers, architects, engineers, executives, or teachers. What do these people look for? What do they need? How can they pay? What does the traditional university have to do to attract and satisfy such very different students? And, finally, what are the wants, needs, values of the elderly? Is there indeed any one "older group," or are there rather several, each with different expectations, needs, values, satisfactions?

Particularly important in age distribution—and with the highest predictive value—are changes in the center of population gravity, that is, in the age group which at any given time constitutes both the largest and the fastest-growing age cohort in the population.

At the end of the Eisenhower presidency, in the late fifties, the center of population gravity in the United States was at its highest point in history. But a violent shift within a few years was bound to take place. As a result of the "baby boom," the center of American population gravity was going to drop so sharply by 1965 as to bring it to the lowest point since the early days of the Republic, to around sixteen or seventeen. It was predictable—and indeed predicted by anyone who took demographics seriously and looked at the figures—that there would be a drastic change in mood and values. The "youth rebellion" of the sixties was mainly a shift of the spotlight to what has always been typical adolescent behavior. In earlier days, with the center of population gravity in the late twenties or early thirties, age groups that are notoriously ultra-conservative, adolescent behavior was dismissed as "Boys will be boys" (and "Girls will be girls"). In the sixties it suddenly became the representative behavior.

But when everybody was talking of a "permanent shift in values" or of a "greening of America," the age pendulum had already swung back, and violently so. By 1969, the first effects of the "baby bust" were already discernible, and not only in the statistics. 1974 or 1975 would be the last year in which the sixteen- and seventeen-year-olds would constitute the center of population gravity. After that, the center would rapidly move up: by the early 1980s it would be in the high twenties again. And with this shift would come a change in what would be considered "representative" behavior. The teenagers would, of course, continue to behave like teenagers. But that would again be dismissed as the way teenagers behave rather than as the constitutive values and behavior of society. And so one could predict with near-certainty, for instance (and some of us did predict it), that by the mid-seventies the college campuses would cease to be "activist" and "rebellious," and college students would again be concerned with grades and jobs; but also that the overwhelming majority of the "dropouts" of 1968 would, ten years later, have become the "upward-mobile professionals" concerned with careers, advancement, tax shelters, and stock options.

Segmentation by educational attainment may be equally important; indeed, for some purposes, it may be more important (e.g., selling encyclopedias, continuing professional education, but also vacation travel). Then there is labor force participation and occupational segmentation. Finally there is income distribution, and especially distribution of disposable and discretionary income. What happens, for instance, to the propensity to save in the two-earner family?

Actually, most of the answers are available. They are the stuff of market research. All that is needed is the willingness to ask the questions.

But more than poring over statistics is involved. To be sure, statistics are the starting point. They were what got Melville to ask what opportunities the jump in teenagers offered a fashion retailer, or what got the top management at Sears, Roebuck to look upon Latin America as a potential market. But then the managements of these companies—or the administrators of metropolitan big-city universities such as Pace in New York and Golden Gate in San Francisco—went out into the field to look and listen.

This is literally how Sears, Roebuck decided to go into Latin America. Sears's chairman, Robert E. Wood, read in the early 1950s that Mexico City and São Paulo were expected to outgrow all U.S. cities by the year 1975. This so intrigued him that he went himself to look at the major cities in Latin America. He spent a week in each of them—Mexico City, Guadalajara, Bogotá, Lima, Santiago, Rio, São Paulo—walking around, looking at stores (he was appalled by what he saw), and studying traffic patterns. Then he knew what customers to aim at, what kind of stores to build, where to put the stores, and what merchandise to stock them with.

Similarly, the founders of Club Mediterranée looked at the customers of package tours, talked to them and listened to them, before they built their first vacation resort. And the two young men who turned Melville Shoe from a dowdy, undistinguished shoe chain (one among many) into the fastest-growing popular fashion retailer in America similarly spent weeks and months in shopping centers, looking at customers, listening to them, exploring their values. They studied the way young people shopped, what kind of environment they liked (do teenage boys and girls, for instance, shop in the same place for shoes or do they want to have separate stores?), and what they considered "value" in the merchandise they bought.

Thus, for those genuinely willing to go out into the field, to look and to listen, changing demographics is both a highly productive and a highly dependable innovative opportunity.

8

Source: Changes in Perception

I

"THE GLASS IS HALF FULL"

In mathematics there is no difference between "The glass is half full" and "The glass is half empty." But the meaning of these two statements is totally different, and so are their consequences. If general perception changes from seeing the glass as "half full" to seeing it as "half empty," there are major innovative opportunities.

Here are a few examples of such changes in perception and of the innovative opportunities they opened up—in business, in politics, in education, and elsewhere.

1. All factual evidence shows that the last twenty years, the years since the early 1960s, have been years of unprecedented advance and improvement in the health of Americans. Whether we look at mortality rates for newborn babies or survival rates for the very old, at occurrence of cancers (other than lung cancer) or cure rates for cancer, and so on, all indicators of physical health and functioning have been moving upward at a good clip. And yet the nation is gripped by collective hypochondria. Never before has there been so much concern with health, and so much fear. Suddenly everything seems to cause cancer or degenerative heart disease or premature loss of memory. The glass is clearly "half empty." What we see now are not the great improvements in health and functioning, but that we are as far away from immortality as ever before and have made no progress toward it. In fact, it can be argued that if there is any real deterioration in American health during the last twenty years it lies precisely in the extreme concern with health and fitness, and the obsession with getting old, with losing fitness, with degenerating into long-term illness or senility.

99

Twenty-five years ago, even minor improvements in the nation's health were seen as major steps forward. Now, even major improvements are barely paid attention to.

Whatever the causes for this change in perception, it has created substantial innovative opportunities. It created, for instance, a market for new health-care magazines: one of them, *American Health*, reached a circulation of a million within two years. It created the opportunity for a substantial number of new and innovative businesses to exploit the fear of traditional foods causing irreparable damage. A firm in Boulder, Colorado, named Celestial Seasonings was started by one of the "flower children" of the late sixties picking herbs in the mountains, packaging them, and peddling them on the street. Fifteen years later, Celestial Seasonings was taking in several hundred million dollars in sales each year and was sold for more than $20 million to a very large food-processing company. And there are highly profitable chains of health-food stores. Jogging equipment has also become big business, and the fastest-growing new business in 1983 in the United States was a company making indoor exercise equipment.

2. Traditionally, the way people feed themselves was very largely a matter of income group and class. Ordinary people "ate"; the rich "dined." This perception has changed within the last twenty years. Now the same people both "eat" and "dine." One trend is toward "feeding," which means getting down the necessary means of sustenance, in the easiest and simplest possible way: convenience foods, TV dinners, McDonald's hamburgers or Kentucky Fried Chicken, and so on. But then the same consumers have also become gourmet cooks. TV programs on gourmet cooking are highly popular and achieve high ratings; gourmet cookbooks have become mass-market best-sellers; whole new chains of gourmet food stores have opened. Finally, traditional supermarkets, while doing 90 percent of their business in foods for "feeding," have opened "gourmet boutiques" which in many cases are far more profitable than their ordinary processed-food business. This new perception is by no means confined to the United States. In West Germany, a young woman physician said to me recently: "Wir essen sechs Tage in der Woche, aber einen Tag wollen wir doch richtig speisen (We feed six days, but one day a week we like to dine)." Not so long ago, "essen" was what ordinary people did seven days a week, and "speisen" what the elite, the rich and the aristocracy, did, seven days a week.

3. If anyone around 1960, in the waning days of the Eisenhower

administration and the beginning of the Kennedy presidency, had predicted the gains the American black would make in the next ten or fifteen years, he would have been dismissed as an unrealistic visionary, if not insane. Even predicting half the gains that those ten or fifteen years actually registered for the American black would have been considered hopelessly optimistic. Never in recorded history has there been a greater change in the status of a social group within a shorter time. At the beginning of those years, black participation in higher education beyond high school was around one-fifth that of whites. By the early seventies, it was equal to that of whites and ahead of that of a good many white ethnic groups. The same rate of advance occurred in employment, in incomes, and especially in entrance to professional and managerial occupations. Anyone granted twelve or fifteen years ago an advance look would have considered the "negro problem" in America to be solved, or at least pretty far along the way toward solution.

But what a large part of the American black population actually sees today in the mid-eighties is not that the glass has become "half full" but that it is still "half empty." In fact, frustration, anger, and alienation have increased rather than decreased for a substantial fraction of the American blacks. They do not see the achievements of two-thirds of the blacks who have moved into the middle class, economically and socially, but the failure of the remaining one-third to advance. What they see is not how fast things have been moving, but how much still remains to be done—how slow and how difficult the going still is. The old allies of the American blacks, the white liberals—the labor unions, the Jewish community, or academia—see the advances. They see that the glass has become "half full." This then has led to a basic split between the blacks and the liberal groups which, of course, only makes the blacks feel even more certain that the glass is "half empty."

The white liberal, however, has come to feel that the blacks increasingly are no longer "deprived," no longer entitled to special treatment such as reverse discrimination, no longer in need of special allowances and priority in employment, in promotion, and so on. This became the opportunity for a new kind of black leader, the Reverend Jesse Jackson. Historically, for almost a hundred years—from Booker T. Washington around the turn of the century through Walter White in the New Deal days until Martin Luther King, Jr., during the presidencies of John Kennedy and Lyndon Johnson—a black could become leader of his community only by proving his ability to get the support of white liberals. It was the one way to obtain enough political strength to make significant gains for American blacks. Jesse Jackson saw that the change in perception that now divides American blacks from their old allies and comrades-in-arms, white liberals, is an innovative opportunity to create a totally different kind of black leadership, one based on vocal enmity to the white liberals and even all-out attack on them. In the past, to have sounded as anti-liberal, anti-union, and anti-Jewish as Jackson has done would have been political suicide. Within a few short weeks in 1984, it made Jackson the undisputed leader of the American black community.

4. American feminists today consider the 1930s and 1940s the darkest of dark ages, with women denied any role in society. Factually, nothing could be more absurd. The America of the 1930s and 1940s was dominated by female stars of the first magnitude. There was Eleanor Roosevelt, the first wife of an American President to establish for herself a major role as a conscience, and as the voice of principle and of compassion which no American male in our history has equaled. Her friend, Frances Perkins, was the first woman in an American cabinet as Secretary of Labor, and the strongest, most effective member of President Roosevelt's cabinet altogether. Anna Rosenberg was the first woman to become a senior executive of a very big corporation as personnel vice-president of R. H. Macy, then the country's biggest retailer; and later on, during the Korean War, she became Assistant Secretary of Defense for manpower and the "boss" of the generals. There were any number of prominent and strong women as university and college presidents, each a national figure. The leading playwrights, Clare Booth Luce and Lillian Hellman, were both women—and Clare Luce then became a major political figure, a member of Congress from Connecticut, and ambassador to Italy. The most publicized medical advance of the period was the work of a woman. Helen Taussig developed the first successful surgery of the living heart, the "blue baby" operation, which saved countless children all over the world and ushered in the age of cardiac surgery, leading directly to the heart transplant and the by-pass operation. And there was Marian Anderson, the black singer and the first black to enter every American living room through the radio, touching the hearts and consciences of millions of Americans as no black before her had done and none would do again until Martin Luther King, Jr., a quarter century later. The list could be continued indefinitely.

These were very proud women, conscious of their achievements, their prominence, their importance. Yet they did not see themselves as "role models." They saw themselves not as women but as individuals. They did not consider themselves as "representative" but as exceptional.

How the change occurred, and why, I leave to future historians to explain. But when it happened around 1970, these great women leaders became in effect "non-persons" for their feminist successors. Now the woman who is not in the labor force, and not working in an occupation traditionally considered "male," is seen as unrepresentative and as the exception.

This was noted as an opportunity by a few businesses, in particular, Citibank (cf. Chapter 7). It was not seen at all, however, by the very industries in which women had long been accepted as professionals and executives, such as department stores, advertising agencies, magazine or book publishers. These traditional employers of professional and managerial women actually today have fewer women in major positions than they had thirty or forty years ago. Citibank, by contrast, was exceedingly *macho*—which may be one reason why it realized there had been a change. It saw in the new perception women had of themselves a major opportunity to court exceptionally able, exceptionally ambitious, exceptionally striving women; to recruit them; and to hold them. And it could do so without competition from the traditional recruiters of career women. In exploiting a change in perception, innovators, as we have seen, can usually count on having the field to themselves for quite a long time.

5. A much older case, one from the early 1950s, shows a similar exploitation of a change in perception. Around 1950, the American population began to describe itself overwhelmingly as being "middle-class," and to do so regardless, almost, of income or occupation. Clearly, Americans had changed their perception of their own social position. But what did the change mean? One advertising executive, William Benton (later senator from Connecticut), went out and asked people what the words "middle class" meant to them. The results were unambiguous: "middle class" in contrast to "working class" means believing in the ability of one's children to rise through performance in school. Benton thereupon bought up the *Encyclopedia Britannica* company and started peddling the *Encyclopedia,* mostly through high school teachers, to parents whose children were the first generation in the family to attend high school. "If you want to be "middle-class," the

salesman said in effect, "your child has to have the *Encyclopedia Britannica* to do well in school." Within three years Benton had turned the almost-dying company around. And ten years later the company began to apply exactly the same strategy in Japan for the same reasons and with the same success.

6. Unexpected success or unexpected failure is often an indication of a change in perception and meaning. Chapter 3 told how the phoenix of the Thunderbird rose from the ashes of the Edsel. What the Ford Motor Company found when it searched for an explanation of the failure of the Edsel was a change in perception. The automobile market, which only a few short years earlier had been segmented by income groups, was now seen by the customers as segmented by "lifestyles."

When a change in perception takes place, the facts do not change. Their meaning does. The meaning changes from "The glass is half full" to "The glass is half empty." The meaning changes from seeing oneself as "working-class" and therefore born into one's "station in life," to seeing oneself as "middle-class" and therefore very much in command of one's social position and economic opportunities. This change can come very fast. It probably did not take much longer than a decade for the majority of the American population to change from considering themselves "working-class" to considering themselves "middle-class."

Economics do not necessarily dictate such changes; in fact, they may be irrelevant. In terms of income distribution, Great Britain is a more egalitarian country than the United States. And yet almost 70 percent of the British population still consider themselves "working-class," even though at least two-thirds of the British population are above "working-class" income by economic criteria alone, and close to half are above the "lower middle class" as well. What determines whether the glass is "half full" or "half empty" is mood rather than facts. It results from experiences that might be called "existential." That the American blacks feel "The glass is half empty" has as much to do with unhealed wounds of past centuries as with anything in present American society. That a majority of the English feel themselves to be "working-class" is still largely a legacy of the nineteenth-century chasm between "church" and "chapel." And the American health hypochondria expresses far more American values, such as the worship of youth, than anything in the health statistics.

Whether sociologists or economists can explain the perceptional phenomenon is irrelevant. It remains a fact. Very often it cannot be quantified; or rather, by the time it can be quantified, it is too late to serve as an opportunity for innovation. But it is not exotic or intangible. It is concrete: it can be defined, tested, and above all exploited.

II

THE PROBLEM OF TIMING

Executives and administrators admit the potency of perception-based innovation. But they tend to shy away from it as "not practical." They consider the perception-based innovator as weird or just a crackpot. But there is nothing weird about the *Encyclopedia Britannica*, about the Ford Thunderbird or Celestial Seasonings. Of course, successful innovators in any field tend to be close to the field in which they innovate. But the only thing that sets them apart is their being alert to opportunity.

One of the foremost of today's gourmet magazines was launched by a young man who started out as food editor of an airlines magazine. He became alert to the change in perception when he read in the same issue of a Sunday paper three contradictory stories. The first said that prepared meals such as frozen dinners, TV dinners, and Kentucky Fried Chicken accounted for more than half of all meals consumed in the United States and were expected to account for three-quarters within a few years. The second said that a TV program on gourmet cooking was receiving one of the highest audience ratings. And the third that a gourmet cookbook in its paperback edition, that is, an edition for the masses, had mounted to the top of the best-seller lists. These apparent contradictions made him ask, What's going on here? A year later he started a gourmet magazine quite different from any that had been on the market before.

Citibank became conscious of the opportunity offered by the moving of women into the work force when its college recruiters reported that they could no longer carry out their instructions, which were to hire the best male business school students in finance and marketing. The best students in these fields, they reported, were increasingly women. College recruiters in many other companies, including quite a few banks, told their managements the same story at that time. In response, most of them were urged, "Just try harder to get the top-flight men." At Citibank, top management saw the change as an opportunity and acted on it.

All these examples, however, also show the critical problem in perception-based innovation: timing. If Ford had waited only one year after the fiasco of the Edsel, it might have lost the "lifestyle" market to GM's Pontiac. If Citibank had not been the first one to recruit women MBAs, it would not have become the preferred employer for the best and most ambitious of the young women aiming to make a career in business.

Yet there is nothing more dangerous than to be premature in exploiting a change in perception. In the first place, a good many of what look like changes in perception turn out to be short-lived fads. They are gone within a year or two. And it is not always apparent which is fad and which is true change. The kids playing computer games were a fad. Companies which, like Atari, saw in them a change in perception lasted one or two years—and then became casualties. Their fathers going in for home computers represented a genuine change, however. It is, furthermore, almost impossible to predict what the consequences of such a change in perception will be. One good example are the consequences of the student rebellions in France, Japan, West Germany, and the United States. Everyone in the late 1960s was quite sure that these would have permanent and profound consequences. But what are they? As far as the universities are concerned, the student rebellions seem to have had absolutely no lasting impact. And who would have expected that, fifteen years later, the rebellious students of 1968 would have become the "Yuppies" to whom Senator Hart appealed in the 1984 American primaries, the young, upward-mobile professionals, ultra-materialistic, job conscious, and maneuvering for their next promotion? There are actually far fewer "dropouts" around these days than there used to be—the only difference is that the media pay attention to them. Can the emergence of homosexuals and lesbians into the limelight be explained by the student rebellion? These were certainly not the results the students themselves in 1968, nor any of the observers and pundits of those days, could possibly have predicted.

And yet, timing is of the essence. In exploiting changes in perception, "creative imitation" (described in Chapter 17) does not work. One has to be first. But precisely because it is so uncertain whether a change in perception is a fad or permanent, and what the consequences really are, perception-based innovation has to start small and be very specific.

9

Source: New Knowledge

Knowledge-based innovation is the "super-star" of entrepreneurship. It gets the publicity. It gets the money. It is what people normally mean when they talk of innovation. Of course, not all knowledge-based innovations are important. Some are truly trivial. But amongst the history-making innovations, knowledge-based innovations rank high. The knowledge, however, is not necessarily scientific or technical. Social innovations based on knowledge can have equal or even greater impact.

Knowledge-based innovation differs from all other innovations in its basic characteristics: time span, casualty rate, predictability, and in the challenges it poses to the entrepreneur. And like most "super-stars," knowledge-based innovation is temperamental, capricious, and hard to manage.

I

THE CHARACTERISTICS OF
KNOWLEDGE-BASED INNOVATION

Knowledge-based innovation has the longest lead time of all innovations. There is, first, a long time span between the emergence of new knowledge and its becoming applicable to technology. And then there is another long period before the new technology turns into products, processes, or services in the marketplace.

Between 1907 and 1910, the biochemist Paul Ehrlich developed the theory of chemotherapy, the control of bacterial microorganisms through chemical compounds. He himself developed the first antibacterial drug, Salvarsan, for the control of syphilis. The sulfa drugs which are the application of Ehrlich's chemotherapy to the control of a broad spectrum of bacterial diseases came on the market after 1936, twenty-five years later.

Rudolph Diesel designed the engine which bears his name in 1897. Everyone at once realized that it was a major innovation. Yet for many years there were few practical applications. Then in 1935 an American, Charles Kettering, totally redesigned Diesel's engine, rendering it capable of being used as the propulsion unit in a wide variety of ships, in locomotives, in trucks, buses, and passenger cars.

A number of knowledges came together to make possible the computer. The earliest was the binary theorem, a mathematical theory going back to the seventeenth century that enables all numbers to be expressed by two numbers only: one and zero. It was applied to a calculating machine by Charles Babbage in the first half of the nineteenth century. In 1890, Hermann Hollerith invented the punchcard, going back to an invention by the early nineteenth-century Frenchman J-M. Jacquard. The punchcard makes it possible to convert numbers into "instructions." In 1906 an American, Lee de Forest, invented the audion tube, and with it created electronics. Then, between 1910 and 1913, Bertrand Russell and Alfred North Whitehead, in their *Principia Mathematica,* created symbolic logic, which enables us to express all logical concepts as numbers. Finally, during World War I, the concepts of programming and feedback were developed, primarily for the purposes of antiaircraft gunnery. By 1918, in other words, all the knowledge needed to develop the computer was available. The first computer became operational in 1946.

A Ford Motor Company manufacturing executive coined the word "automation" in 1951 and described in detail the entire manufacturing process automation would require. "Robotics" and factory automation were widely talked about for twenty-five years, but nothing really happened for a long time. Nissan and Toyota in Japan did not introduce robots into their plants until 1978. In the early eighties, General Electric built an automated locomotive plant in Erie, Pennsylvania. General Motors then began to automate several of its engine and accessory plants. Early in 1985, Volkswagen began to operate its "Hall 54" as an almost completely automated manufacturing installation.

Buckminster Fuller, who called himself a geometer and who was part mathematician and part philosopher, applied the mathematics of topology to the design of what he called the "Dymaxion House," a term he chose because he liked the sound of it. The Dymaxion House combines the greatest possible living space with the smallest possible surface. It therefore has optimal insulation, optimal heating and cooling, and superb acoustics. It also can be built with lightweight materials, requires no foundation and a minimum of suspension, and can still withstand an earthquake or the fiercest gale. Around 1940, Fuller put a Dymaxion House on the campus of a small New England college. And there it stayed. Very few Dymaxion Houses have been built—Americans, it seems, do not like to live in circular homes. But around 1965, Dymaxion structures began to be put up in the Arctic and Antarctic where conventional buildings are impractical, expensive, and difficult to erect. Since then they have increasingly been used for large structures such as auditoriums, concert tents, sports arenas, and so on.

Only major external crises can shorten this lead time. De Forest's audion tube, invented in 1906, would have made radio possible almost immediately, but it would still not have been on the market until the late 1930s or so had not World War I forced governments, and especially the American government, to push the development of wireless transmission of sounds. Field telephones connected by wires were simply too unreliable, and wireless telegraphy was confined to dots and dashes. And so, radio came on the market early in the 1920s, only fifteen years after the emergence of the knowledge on which it is based.

Similarly, penicillin would probably not have been developed until the 1950s or so but for World War II. Alexander Fleming found the bacteria-killing mold, penicillium, in the mid-twenties. Howard Florey, an English biochemist, began to work on it ten years later. But it was World War II that forced the early introduction of penicillin. The need to have a potent drug to fight infections led the British government to push Florey's research: English soldiers were made available to him as guinea pigs wherever they fought. The computer, too, would probably have waited for the discovery of the transistor by Bell Lab physicists in 1947 had not World War II led the American government to push computer research and to invest large resources of men and money in the work.

The long lead time for knowledge-based innovations is by no means confined to science or technology. It applies equally to innovations that are based on nonscientific and nontechnological knowledge.

The comte de Saint-Simon developed the theory of the entrepreneurial bank, the purposeful use of capital to generate economic development, right after the Napoleonic wars. Until then bankers were moneylenders who lent against "security" (e.g., the taxing power of a prince). Saint-Simon's banker was to "invest," that is, to create new wealth-producing capacity. Saint-Simon had extraordinary influence in his time, and a popular cult developed around his memory and his ideas after his death in 1826. Yet it was not until 1852 that two disciples, the brothers Jacob and Isaac Pereire, established the first entrepreneurial bank, the Crédit Mobilier, and with it ushered in what we now call finance capitalism.

Similarly, many of the elements needed for what we now call management were available right after World War I. Indeed, in 1923, Herbert Hoover, soon to be President of the United States, and Thomas Masaryk, founder and president of Czechoslovakia, convened the first International Management Congress in Prague. At the same time a few large companies here and there, especially DuPont and General Motors in the United States, began to reorganize themselves around the new management concepts. In the next decade a few "true believers," especially an Englishman, Lyndall Urwick, the founder of the first management consulting firm which still bears his name, began to write on management. Yet it was not until my *Concept of the Corporation* (1946) and *Practice of Management* (1954) were published that management become a discipline accessible to managers all over the world. Until then each student or practitioner of "management" focused on a separate area; Urwick on organization, others on the management of people, and so on. My books codified it, organized it, systematized it. Within a few years, management became a worldwide force.

Today, we experience a similar lead time in respect to learning theory. The scientific study of learning began around 1890 with Wilhelm Wundt in Germany and William James in the United States. After World War II, two Americans—B. F. Skinner and Jerome Bruner, both at Harvard—developed and tested basic theories of learning, Skinner specializing in behavior and Bruner in cognition. Yet only now is learning theory beginning to become a factor in our schools. Perhaps the time has come for an entrepreneur to start schools based on what we know about learning, rather than on the old wives' tales about it that have been handed down through the ages.

In other words, the lead time for knowledge to become applicable technology and begin to be accepted on the market is between twenty-five and thirty-five years.

This has not changed much throughout recorded history. It is widely believed that scientific discoveries turn much faster in our day than ever before into technology, products, and processes. But this is largely illusion. Around 1250 the Englishman Roger Bacon, a Franciscan monk, showed that refraction defects of the eye could be corrected with eyeglasses. This was incompatible with what everybody then knew: the "infallible" authority of the Middle Ages Galen, the great medical scientist, had "proven conclusively" that it could not be done. Roger Bacon lived and worked on the extreme edges of the civilized world, in the wilds of northern Yorkshire. Yet a mural, painted thirty years later in the Palace of the Popes in Avignon (where it can still be seen), shows elderly cardinals wearing reading glasses; and ten years later, miniatures show elderly courtiers in the Sultan's Palace in Cairo also in glasses. The mill race, which was the first true "automation," was developed to grind grain by the Benedictine monks in northern Europe around the year 1000; within thirty years it had spread all over Europe. Gutenberg's invention of movable type and the woodcut both followed within thirty years of the West's learning of Chinese printing.

The lead time for knowledge to become knowledge-based innovation seems to be inherent in the nature of knowledge. We do not know why. But perhaps it is not pure coincidence that the same lead time applies to new scientific theory. Thomas Kuhn, in his path-breaking book *The Structure of Scientific Revolutions* (1962), showed that it takes about thirty years before a new scientific theory becomes a new paradigm—a new statement that scientists pay attention to and use in their own work.

CONVERGENCES

The second characteristic of knowledge-based innovations—and a truly unique one—is that they are almost never based on one factor but on the convergence of several different kinds of knowledge, not all of them scientific or technological.

Few knowledge-based innovations in this century have benefited humanity more than the hybridization of seeds and livestock. It enables the earth to feed a much larger population than anyone would have thought possible fifty years ago. The first successful new seed was hybrid corn. It was produced after twenty years of hard work by Henry C.

Wallace, the publisher of a farm newspaper in Iowa, and later U.S. Secretary of Agriculture under Harding and Coolidge—the only holder of this office, perhaps, who deserves to be remembered for anything other than giving away money. Hybrid corn has two knowledge roots. One was the work of the Michigan plant breeder William J. Beal, who around 1880 discovered hybrid vigor. The other was the rediscovery of Mendel's genetics by the Dutch biologist Hugo de Vries. The two men did not know of one another. Their work was totally different both in intent and content. But only by pulling it together could hybrid corn be developed.

The Wright Brothers' airplane also had two knowledge roots. One was the gasoline engine, designed in the mid-1880s to power the first automobiles built by Karl Benz and Gottfried Daimler, respectively. The other one was mathematical: aerodynamics, developed primarily in experiments with gliders. Each was developed quite independently. It was only when the two came together that the airplane became possible.

The computer, as already noted, required the convergence of no less than five different knowledges: a scientific invention, the audion tube; a major mathematical discovery, the binary theorem; a new logic; the design concept of the punchcard; and the concepts of program and feedback. Until all these were available, no computer could have been built. Charles Babbage, the English mathematician, is often called the "father of the computer." What kept Babbage from building a computer, it is argued, was only the unavailability of the proper metals and of electric power at his time. But this is a misunderstanding. Even if Babbage had had the proper materials, he could at best have built the mechanical calculator that we now call a cash register. Without the logic, the design concept of the punchcard, and the concept of program and feedback, none of which Babbage possessed, he could only imagine a computer.

The Brothers Pereire founded the first entrepreneurial bank in 1852. It failed within a few years because they had only one knowledge base and the entrepreneurial bank needs two. They had a theory of creative finance that enabled them to be brilliant venture capitalists. But they lacked the systematic knowledge of banking which was developed at exactly the same time across the Channel by the British, and codified in Walter Bagehot's classic, *Lombard Street*.

After their failure in the early 1860s, three young men independently picked up where the Brothers Pereire had left off, added the knowledge base of banking to the venture capital concept, and succeeded. The first was J. P. Morgan, who had been trained in London but had also carefully studied the Pereires' Crédit Mobilier. He founded the most successful entrepreneurial bank of the nineteenth century in New York in 1865. The second one, across the Rhine, was the young German Georg Siemens, who founded what he called the "Universal Bank," by which he meant a bank that was both a deposit bank on the British model and an entrepreneurial bank on the Pereires' model. And in remote Tokyo, another young man, Shibusawa Eichii, who had been one of the first Japanese to travel to Europe to study banking first-hand, and had spent time both in Paris and in London's Lombard Street, became one of the founders of the modern Japanese economy by establishing a Japanese version of the Universal Bank. Both Siemens's Deutsche Bank and Shibusawa's Daichi Bank are still the largest banks of their respective countries.

The first man to envisage the modern newspaper was an American, James Gordon Bennett, who founded the *New York Herald*. Bennett fully understood the problems: A newspaper had to have enough income to be editorially independent and yet be cheap enough to have mass circulation. Earlier newspapers either got their income by selling their independence and becoming the lackeys and paid propagandists of a political faction—as did most American and practically all European papers of his time. Or, like the great aristocrat of those days, *The Times* of London, they were "written by gentlemen for gentlemen," but so expensive that only a small elite could afford them.

Bennett brilliantly exploited the twin technological knowledge bases on which a modern newspaper rests: the telegraph and high-speed printing. They enabled him to produce a paper at a fraction of the traditional cost. He knew that he needed high-speed typesetting, though it was not invented until after his death. He also saw one of the two nonscientific bases, mass literacy, which made possible mass circulation for a cheap newspaper. But he failed to grasp the fifth base: mass advertising as the source of the income that makes possible editorial independence. Bennett personally enjoyed a spectacular success; he was the first of the press lords. But his newspaper achieved neither leadership nor financial security. These goals were only attained two decades later, around 1890, by three men who understood and exploited advertising: Joseph Pulitzer, first in St. Louis and then in New

York; Adolph Ochs, who took over a moribund *New York Times* and made it into America's leading paper; and William Randolph Hearst, who invented the modern newspaper chain.

The invention of plastics, beginning with Nylon, also rested on the convergence of a number of different new knowledges each emerging around 1910. Organic chemistry, pioneered by the Germans and perfected by Leo Baekeland, a Belgian working in New York, was one; X-Ray diffraction and with it an understanding of the structure of crystals was another; and high-vacuum technology. The final factor was the pressure of World War I shortages, which made the German government willing to invest heavily in polymerization research to obtain a substitute for rubber. It took a further twenty years, though, before Nylon was ready for the market.

Until all the needed knowledges can be provided, knowledge-based innovation is premature and will fail. In most cases, the innovation occurs only when these various factors are already known, already available, already in use someplace. This was the case with the Universal Bank of 1865–75. It was the case with the computer after World War II. Sometimes the innovator can identify the missing pieces and then work at producing them. Joseph Pulitzer, Adolph Ochs, and William Randolph Hearst largely created modern advertising. This then created what we today call media, that is, the merger of information and advertising in "mass communications." The Wright Brothers identified the pieces of knowledge that were missing—mostly mathematics—and then themselves developed them by building a wind tunnel and actually testing mathematical theories. But until all the knowledges needed for a given knowledge-based innovation have come together, the innovation will not take off. It will remain stillborn.

Samuel Langley, for instance, whom his contemporaries expected to become the inventor of the airplane, was a much better trained scientist than the Wright Brothers. As secretary of what was then America's leading scientific institution, the Smithsonian in Washington, he also had all the nation's scientific resources at his disposal. But even though the gasoline engine had been invented by Langley's time, he preferred to ignore it. He believed in the steam engine. As a result his airplane could fly; but because of the steam engine's weight, it could not carry any load, let alone a pilot. It needed the convergence of mathematics and the gasoline engine to produce the airplane.

Indeed, until all the knowledges converge, the lead time of a knowledge-based innovation usually does not even begin.

II

WHAT KNOWLEDGE-BASED INNOVATION
REQUIRES

Its characteristics give knowledge-based innovation specific requirements. And these requirements differ from those of any other kind of innovation.

1. In the first place, knowledge-based innovation requires careful analysis of all the necessary factors, whether knowledge itself, or social, economic, or perceptual factors. The analysis must identify what factors are not yet available so that the entrepreneur can decide whether these missing factors can be produced—as the Wright Brothers decided in respect to the missing mathematics—or whether the innovation had better be postponed as not yet feasible.

The Wright Brothers exemplify the method at its best. They thought through carefully what knowledge was necessary to build an airplane for manned, motored flight. Next they set about to develop the pieces of knowledge that were needed, taking the available information, testing it first theoretically, then in the wind tunnel, and then in actual flight experiments, until they had the mathematics they needed to construct ailerons, to shape the wings, and so on.

The same analysis is needed for nontechnical knowledge-based innovation. Neither J. P. Morgan nor Georg Siemens published their papers; but Shibusawa in Japan did. And so we know that he based his decision to forsake a brilliant government career and to start a bank on a careful analysis of the knowledge available and the knowledge needed. Similarly, Joseph Pulitzer analyzed carefully the knowledge needed when he launched what became the first modern newspaper, and decided that advertising had to be invented and could be invented.

If I may inject a personal note, my own success as an innovator in the management field was based on a similar analysis in the early 1940s. Many of the required pieces of knowledge were already available: organization theory, for instance, but also quite a bit of knowledge about managing work and worker. My analysis also showed, however, that these pieces were scattered and lodged in half a dozen different disciplines. Then it found which key knowledges were missing: purpose of a business; any knowledge of the work and structure of top management; what we now term "business policy" and "strategy"; objectives; and so on. All of the missing knowledges, I decided, could be produced. But without such analysis, I could never have known what they were or that they were missing.

Failure to make such an analysis is an almost sure-fire prescription for disaster. Either the knowledge-based innovation is not achieved, which is what happened to Samuel Langley. Or the innovator loses the fruits of his innovation and only succeeds in creating an opportunity for somebody else.

Particularly instructive is the failure of the British to reap the harvest from their own knowledge-based innovations.

The British discovered and developed penicillin, but it was the Americans who took it over. The British scientists did a magnificent technical job. They came out with the right substances and the right uses. Yet they failed to identify the ability to manufacture the stuff as a critical knowledge factor. They could have developed the necessary knowledge of fermentation technology; they did not even try. As a result, a small American company, Pfizer, went to work on developing the knowledge of fermentation and became the world's foremost manufacturer of penicillin.

Similarly, the British conceived, designed, and built the first passenger jet plane. But de Havilland, the British company, did not analyze what was needed and therefore did not identify two key factors. One was configuration, that is, the right size with the right payload for the routes on which the jet would give an airline the greatest advantage. The other was equally mundane: how to finance the purchase of such an expensive plane by the airlines. As a result of de Havilland's failure to do the analysis, two American companies, Boeing and Douglas, took over the jet plane. And de Havilland has long since disappeared.

Such analysis would appear to be fairly obvious, yet it is rarely done by the scientific or technical innovator. Scientists and technologists are reluctant to make these analyses precisely because they think they already *know*. This explains why, in so many cases, the great knowledge-based innovations have had a layman rather than a scientist or a technologist for their father, or at least their godfather. The (American) General Electric Company is largely the brainchild of a financial man.

He conceived the strategy (described in Chapter 19) that made G.E. the world's leading supplier of large steam turbines and, therewith, the world's leading supplier to electric power producers. Similarly, two laymen, Thomas Watson, Sr., and his son Thomas Watson, Jr., made IBM the leader in computers. At DuPont, the analysis of what was needed to make the knowledge-based innovation of Nylon effective and successful was not done by the chemist who developed the technology, but by business people on the executive committee. And Boeing became the world's leading producer of jet planes under the leadership of marketing people who understood what the airlines and the public needed.

This is not a law of nature, however. Mostly it is a matter of will and self-discipline. There have been plenty of scientists and technologists—Edison is a good example—who forced themselves to think through what their knowledge-based innovation required.

2. The second requirement of knowledge-based innovation is a clear focus on the strategic position. It cannot be introduced tentatively. The fact that the introduction of the innovation creates excitement, and attracts a host of others, means that the innovator has to be right the first time. He is unlikely to get a second chance. In all the other innovations discussed so far, the innovator, once he has been successful with his innovation, can expect to be left alone for quite some time. This is not true of knowledge-based innovation. Here the innovators almost immediately have far more company than they want. They need only stumble once to be overrun.

There are basically only three major focuses for knowledge-based innovation. First, there is the focus Edwin Land took with Polaroid: To develop a *complete system* that would then dominate the field. This is exactly what IBM did in its early years when it chose not to sell computers but to lease them to its customers. It supplied them with such software as was available, with programming, with instruction in computer language for programmers, with instruction in computer use for a customer's executives, and with service. This was also what G.E. did when it established itself as the leader in the knowledge-based innovation of large steam turbines in the early years of this century.

The second clear focus is a *market focus.* Knowledge-based innovation can aim at creating the market for its products. This is what DuPont did with Nylon. It did not "sell" Nylon; it created a consumer market for women's hosiery and women's underwear using Nylon, a market for automobile tires using Nylon, and so on. It then delivered Nylon to the fabricators to make the articles for which DuPont had already created a demand and which, in effect, it had already sold. Similarly, aluminum from the very beginning, right after the invention of the aluminum reduction process by Charles M. Hall in 1888, began to create a market for pots and pans, for rods and other aluminum extrusions. The aluminum company actually went into making these end products and selling them. It created the market which, in turn, discouraged (if it did not keep out altogether) potential competitors.

The third focus is *to occupy a strategic position*, concentrating on a key function (the strategy is discussed in Chapter 18 under Ecological Niches). What position would enable the knowledge innovator to be largely immune to the extreme convolutions of a knowledge-based industry in its early stages? It was thinking this through and deciding to concentrate on mastering the fermentation process that gave Pfizer in the United States the early lead in penicillin it has maintained ever since. Focusing on marketing—on mastery of the requirements of airlines and of the public in respect to configuration and finance—gave Boeing the leadership in passenger planes, which it has held ever since. And despite the turbulence of the computer industry today, a few leading manufacturers of the computer's key component, semiconductors, can maintain their leadership position almost irrespective of the fate of individual computer manufacturers themselves. Intel is one example.

Within the same industry, individual knowledge-based innovators can sometimes choose between these alternatives. Where DuPont, for instance, has chosen to create markets, its closest American competitor, Dow Chemical, tries to occupy a key spot in each market segment. A hundred years ago, J. P. Morgan opted for the key function approach. He established his bank as the conduit for European investment capital in American industry, and furthermore in a capital-short country. At the same time, Georg Siemens in Germany and Shibusawa Eichii in Japan both went for the systems approach.

The power of a clear focus is demonstrated by Edison's success. Edison was not the only one who identified the inventions that had to be made to produce a light bulb. An English physicist, Joseph Swan, did so too. Swan developed his light bulb at exactly the same time as Edison. Technically, Swan's bulb was superior, to the point where Edison bought up the Swan patents and used them in his own light bulb factories. But Edison not only thought through the technical requirements;

he thought through his focus. Before he even began the technical work on the glass envelope, the vacuum, the closure, and the glowing fiber, he had already decided on a "system": his light bulb was designed to fit an electric power company for which he had lined up the financing, the rights to string wires to get the power to his light bulb customers, and the distribution system. Swan, the scientist, invented a product; Edison produced an industry. So Edison could sell and install electric power while Swan was still trying to figure out who might be interested in his technical achievement.

The knowledge-based innovator has to decide on a clear focus. Each of the three described here is admittedly very risky. But not to decide on a clear focus, let alone to try to be in between or to attempt more than one focus, is riskier by far. It is likely to prove fatal.

3. Finally, the knowledge-based innovator—and especially the one whose innovation is based on scientific or technological knowledge—needs to learn and to practice entrepreneurial management (see Chapter 15, The New Venture). In fact, entrepreneurial management is more crucial to knowledge-based innovation than to any other kind. Its risks are high, thus putting a much higher premium on foresight, both financial and managerial, and on being market-focused and market-driven. Yet knowledge-based, and especially high-tech, innovation tends to have little entrepreneurial management. In large measure the high casualty rate of knowledge-based industry is the fault of the knowledge-based, and especially the high-tech, entrepreneurs themselves. They tend to be contemptuous of anything that is not "advanced knowledge," and particularly of anyone who is not a specialist in their own area. They tend to be infatuated with their own technology, often believing that "quality" means what is technically sophisticated rather than what gives value to the user. In this respect they are still, by and large, nineteenth-century inventors rather than twentieth-century entrepreneurs.

In fact, there are enough companies around today to show that the risk in knowledge-based innovation, including high tech, can be substantially reduced if entrepreneurial management is conscientiously applied. Hoffmann-LaRoche, the Swiss pharmaceutical company, is one example; Hewlett-Packard is another, and so is Intel. Precisely because the inherent risks of knowledge-based innovation are so high, entrepreneurial management is both particularly necessary and particularly effective.

III

Even when it is based on meticulous analysis, endowed with clear focus, and conscientiously managed, knowledge-based innovation still suffers from unique risks and, worse, an innate unpredictability.

First, by its very nature, it is turbulent.

The combination of the two characteristics of knowledge-based innovations—long lead times and convergences—gives knowledge-based innovations their peculiar rhythm. For a long time, there is awareness of an innovation about to happen—but it does not happen. Then suddenly there is a near-explosion, followed by a few short years of tremendous excitement, tremendous startup activity, tremendous publicity. Five years later comes a "shakeout," which few survive.

In 1856, Werner Siemens in Germany applied the electrical theories Michael Faraday had developed around 1830 (twenty-five years earlier) to the design of the ancestor of the first electrical motor, the first dynamo. It caused a worldwide sensation. From then on, it became certain that there would be an "electrical industry" and that it would be a major one. Dozens of scientists and inventors went to work. But nothing happened for twenty-two years. The knowledge was missing: Maxwell's development of Faraday's theories.

After it had become available, Edison invented the light bulb in 1878 and the race was on. Within the next five years all the major electrical apparatus companies in Europe and America were founded: Siemens in Germany bought up a small electrical apparatus manufacturer, Schuckert. The (German) General Electric Company, AEG, was formed on the basis of Edison's work. In the United States there arose what are now G.E. and Westinghouse; in Switzerland, there was Brown Boveri; in Sweden, ASEA was founded in 1884. But these few are the survivors of a hundred such companies—American, British, French, German, Italian, Spanish, Dutch, Belgian, Swiss, Austrian, Czech, Hungarian, and so on—all eagerly financed by the investors of their time and all expecting to be "billion-dollar companies." It was this upsurge of the electrical apparatus industry that gave rise to the first great science-fiction boom and made Jules Verne and H. G. Wells best-selling authors all over the world. But by 1895–1900, most of these companies had already disappeared, whether out of business, bankrupt, or absorbed by the few survivors.

Around 1910, there were up to two hundred automobile companies in the United States alone. By the early 1930s, their number had shrunk to twenty, and by 1960 to four.

In the 1920s, literally hundreds of companies were making radio sets and hundreds more were going into radio stations. By 1935, the control of broadcasting had moved into the hands of three "networks" and there were only a dozen manufacturers of radio sets left. Again, there was an explosion in the number of newspapers founded between 1880 and 1900. In fact, newspapers were among the major "growth industries" of the time. Since World War I, the number of newspapers in every major country has been going downhill steadily. And the same is true of banking. After the founders—the Morgans, the Siemenses, the Shibusawas—there was an almost explosive growth of new banks in the United States as well as in Europe. But around 1890, only twenty years later, consolidation set in. Banking firms began to go out of business or to merge. By the end of World War II in every major country only a handful of banks were left that had more than local importance, whether as commercial or private banks.

But each time without exception the survivor has been a company that was started during the early explosive period. After that period is over, entry into the industry is foreclosed for all practical purposes. There is a "window" of a few years during which a new venture must establish itself in any new knowledge-based industry.

It is commonly believed today that that "window" has become narrower. But this is as much a misconception as the common belief that the lead time between the emergence of new knowledge and its conversion into technology, products, and processes has become much shorter.

Within a few years after George Stephenson's "Rocket" had pulled the first train on a commercial railroad in 1830, over a hundred railroad companies were started in England. For ten years railroads were "high-tech" and railroad entrepreneurs "media events." The speculative fever of these years is bitingly satirized in one of Dickens's novels, *Little Dorrit* (published in 1855–57); it was not very different from today's speculative fever in Silicon Valley. But around 1845, the "window" slammed shut. From then on there was no money in England any more for new railroads. Fifty years later, the hundred-or-so English railroad companies of 1845 had shrunk to five or six. And the same rhythm characterized the electrical apparatus industry, the telephone industry, the automobile industry, the chemical industry, household appliances, and consumer electronics. The "window" has never been very wide nor open very long.

But there can be little doubt that today the "window" is becoming more and more crowded. The railroad boom of the 1830s was confined to England; later, every country had its own local boom quite separate from the preceding one in the neighboring country. The electrical apparatus boom already extended across national frontiers, as did the automobile boom twenty-five years later. Yet both were confined to the countries that were industrially developed at the time. The term "industrially developed" encompasses a great deal more territory today, however. It takes in Japan, for instance. It takes in Brazil. It may soon take in the non-Communist Chinese territories: Hong Kong, Taiwan, and Singapore. Communication today is practically instantaneous, travel easy and fast. And a great many countries have today what only very few small places had a hundred years ago: large cadres of trained people who can immediately go to work in any area of knowledge-based innovation, and especially of science-based or technology-based innovation.

These facts have two important implications.

1. First, science-based and technology-based innovators alike find time working against them. In all innovation based on any other source —the unexpected, incongruities, process need, changes in industry structure, demographics, or changes in perception—time is on the side of the innovator. In any other kind of innovation innovators can reasonably expect to be left alone. If they make a mistake, they are likely to have time to correct it. And there are several moments in time in which they can launch their new venture. Not so in knowledge-based innovation, and especially in those innovations based on scientific and technological knowledge. Here there is only a short time—the "window"— during which entry is possible at all. Here innovators do not get a second chance; they have to be right the first time. The environment is harsh and unforgiving. And once the "window" closes, the opportunity is gone forever.

In some knowledge-based industries, however, a second "window" does in fact open some twenty to thirty years or so after the first one has shut down. Computers are an example.

The first "window" in computers lasted from 1949 until 1955 or so. During this period, every single electrical apparatus company in the world went into computers—G.E., Westinghouse, and RCA in the United States; the British General Electric Company, Plessey, and Ferranti in Great Britain; Siemens and AEG in Germany; Philips in Holland; and so on. By 1970, every single one of the "biggies" was out of computers, ignominiously. The field was occupied by companies that had either not existed at all in 1949 or had been small and marginal: IBM, of course, and the "Seven Dwarfs," the seven smaller computer companies in the United States; ICL, the remnant of the computer businesses of the General Electric Company, of Plessey, and of Ferranti in Great Britain; some fragments sustained by heavy government subsidies in France; and a total newcomer, Nixdorf, in Germany. The Japanese companies were sustained for a long time through government support.

Then, in the late seventies, a second "window" opened with the invention of micro-chips, which led to word processors, minicomputers, personal computers, and the merging of computer and telephone switchboard.

But the companies that had failed in the first round did not come back in the second one. Even those that survived the first round stayed out of the second, or came in late and reluctantly. Neither Univac nor Control Data, nor Honeywell nor Burroughs, nor Fujitsu nor Hitachi took leadership in minicomputers or personal computers. The one exception was IBM, the undisputed champion of the first round. And this has been the pattern too in earlier knowledge-based innovations.

2. Because the "window" is much more crowded, any one knowledge-based innovator has far less chance of survival.

The number of entrants during the "window" period is likely to be much larger. But the structure of the industries, once they stabilize and mature, seems to have remained remarkably unchanged, at least for a century now. Of course there are great differences in structure between various industries, depending on technology, capital requirements, and ease of entry, on whether the product can be shipped or distributed only locally, and so on. But at any one time any given industry has a typical structure: in any given market there are so many companies altogether, so many big ones, so many medium-sized ones, so many small ones, so many specialists. And increasingly there is only one "market" for any new knowledge-based industry, whether computers or modern banking—the world market.

The number of knowledge-based innovators that will survive when an industry matures and stabilizes is therefore no larger than it has traditionally been. But largely because of the emergence of a world market and of global communications, the number of entrants during the "window" period has greatly increased. When the shakeout comes, the casualty rate is therefore much higher than it used to be. And the shakeout always comes; it is inevitable.

THE SHAKEOUT

The "shakeout" sets in as soon as the "window" closes. And the majority of ventures started during the "window" period do not survive the shakeout, as has already been shown for such high-tech industries of yesterday as railroads, electrical apparatus makers, and automobiles. As these lines are being written, the shakeout has begun among microprocessor, minicomputer, and personal computer companies—only five or six years after the "window" opened. Today, there are perhaps a hundred companies in the industry in the United States alone. Ten years hence, by 1995, there are unlikely to be more than a dozen left of any size or significance.

But which ones will survive, which ones will die, and which ones will become permanently crippled—able neither to live nor to die—is unpredictable. In fact, it is futile to speculate. Sheer size may ensure survival. But it does not guarantee success in the shakeout, otherwise Allied Chemical rather than DuPont would today be the world's biggest and most successful chemical company. In 1920, when the "window" opened for the chemical industry in the United States, Allied Chemical looked invincible, if only because it had obtained the German chemical patents which the U.S. government had confiscated during World War I. Seven years later, after the shakeout, Allied Chemical had become a weak also-ran. It has never been able to regain momentum.

No one in 1949 could have predicted that IBM would emerge as the computer giant, let alone that such big, experienced leaders as G.E. or Siemens would fail completely. No one in 1910 or 1914 when automobile stocks were the favorites of the New York Stock Exchange could have predicted that General Motors and Ford would survive and prosper and that such universal favorites as Packard or Hupmobile would disappear. No one in the 1870s and 1880s, the period in which the modern banks were born, could have predicted that Deutsche Bank would swallow up dozens of the old commercial banks of Germany and emerge as the leading bank of the country.

That a certain industry will become important is fairly easy to predict. There is no case on record where an industry that reached the explosive phase, the "window" phase, as I called it, has then failed to become a major industry. The question is, Which of the specific units in this industry will be its leaders and so survive?

This rhythm—a period of great excitement during which there is also great speculative ferment, followed by a severe "shakeout"—is particularly pronounced in the high-tech industries.

In the first place, such industries are in the limelight and thus attract far more entrants and far more capital than more mundane areas. Also the expectations are much greater. More people have probably become rich building such prosaic businesses as a shoe-polish or a watchmaking company than have become rich through high-tech businesses. Yet no one expects shoe-polish makers to build a "billion-dollar business," nor considers them a failure if all they build is a sound but modest family company. High tech, by contrast, is a "high–low game," in which a middle hand is considered worthless. And this makes high-tech innovation inherently risky.

But also, high tech is not profitable for a very long time. The world's computer industry began in 1947–48. Not until the early 1980s, more than thirty years later, did the industry as a whole reach break-even point. To be sure, a few companies (practically all of them American, by the way) began to make money much earlier. And one, IBM, the leader, began to make a great deal of money earlier still. But across the industry the profits of those few successful computer makers were more than offset by the horrendous losses of the rest; the enormous losses, for instance, which the big international electrical companies took in their abortive attempts to become computer manufacturers.

And exactly the same thing happened in every earlier "high-tech" boom—in the railroad booms of the early nineteenth century, in the electrical apparatus and the automobile booms between 1880 and 1914, in the electric appliance and the radio booms of the 1920s, and so on.

One major reason for this is the need to plow more and more money back into research, technical development, and technical services to stay in the race. High tech does indeed have to run faster and faster in order to stand still.

This is, of course, part of its fascination. But it also means that when the shakeout comes, very few businesses in the industry have the financial resources to outlast even a short storm. This is the reason why high-tech ventures need financial foresight even more than other new ventures, but also the reason why financial foresight is even scarcer among high-tech new ventures than it is among new ventures in general.

There is only one prescription for survival during the shakeout: entrepreneurial management (described in Chapters 12–15). What distinguished Deutsche Bank from the other "hot" financial institutions of its time was that Georg Siemens thought through and built the world's first top management team. What distinguished DuPont from Allied Chemical was that DuPont in the early twenties created the world's first systematic organization structure, the world's first long-range planning, and the world's first system of management information and control. Allied Chemical, by contrast, was run arbitrarily by one brilliant egomaniac. But this is not the whole story. Most of the large companies that failed to survive the more recent computer shakeout—G.E. and Siemens, for instance—are usually considered to have first-rate management. And the Ford Motor Company survived, though only by the skin of its teeth, even though it was grotesquely mismanaged during the shakeout years.

Entrepreneurial management is thus probably a precondition of survival, but not a guarantee thereof. And at the time of the shakeout, only insiders (and perhaps not even they) can really know whether a knowledge-based innovator that has grown rapidly for a few boom years is well managed, as DuPont was, or basically unmanaged, as Allied Chemical was. By the time we do know, it is likely to be too late.

THE RECEPTIVITY GAMBLE

To be successful, a knowledge-based innovation has to be "ripe"; there has to be receptivity to it. This risk is inherent in knowledge-based innovation and is indeed a function of its unique power. All other innovations exploit a change that has already occurred. They satisfy a need that already exists. But in knowledge-based innovation, the innovation brings about the change. It aims at creating a want. And no one can tell in advance whether the user is going to be receptive, indifferent, or actively resistant.

There are exceptions, to be sure. Whoever produces a cure for cancer need not worry about "receptivity." But such exceptions are few. In most knowledge-based innovations, receptivity is a gamble. And the odds are unknown, are indeed mysterious. There may be great receptivity, yet no one realizes it. And there may be no receptivity, or even heavy resistance when everyone is quite sure that society is actually eagerly waiting for the innovation.

Stories of the obtuseness of the high and mighty in the face of a knowledge-based innovation abound. Typical is the anecdote which has a king of Prussia predicting the certain failure of that new-fangled contraption, the railroad, because "No one will pay good money to get from Berlin to Potsdam in one hour when he can ride his horse in one day for free." But the king of Prussia was not alone in his misreading of the receptivity to the railroad; the majority of the "experts" of his day inclined to his opinion. And when the computer appeared there was not one single "expert" who could imagine that businesses would ever want such a contraption.

The opposite error is, however, just as common. "Everybody knows" that there is a real need, a real demand, when in reality there is total indifference or resistance. The same authorities who, in 1948, could not imagine that a business would ever want a computer, a few years later, around 1955, predicted that the computer would "revolutionize the schools" within a decade.

The Germans consider Philip Reis rather than Alexander Graham Bell to be the inventor of the telephone. Reis did indeed build an instrument in 1861 that could transmit music and was very close to transmitting speech. But then he gave up, totally discouraged. There was no receptivity for a telephone, no interest in it, no desire for it. "The telegraph is good enough for us," was the prevailing attitude. Yet when Bell, fifteen years later, patented his telephone, there was an immediate enthusiastic response. And nowhere was it greater than in Germany.

The change in receptivity in these fifteen years is not too difficult to explain. Two major wars, the American Civil War and the Franco-

Prussian War, had shown that the telegraph was by no means "good enough." But the real point is not why receptivity changed. It is that every authority in 1861 enthusiastically predicted overwhelming receptivity when Reis demonstrated his instrument at a scientific meeting. And every authority was wrong.

But, of course, the authorities can also be right, and often are. In 1876–77, for instance, they all knew that there was receptivity for both a light bulb and a telephone—and they were right. Similarly, Edison, in the 1880s, was supported by the expert opinion of his time when he embarked on the invention of the phonograph, and again the experts were right in assuming high receptivity for the new device.

But only hindsight can tell us whether the experts are right or wrong in their assessment of the receptivity for this or that knowledge-based innovation.

Nor do we necessarily perceive, even by hindsight, why a particular knowledge-based innovation has receptivity or fails to find it. No one, for instance, can explain why phonetic spelling has been so strenuously resisted. Everyone agrees that nonphonetic spelling is a major obstacle in learning to read and write, forces schools to devote inordinate time to the reading skill, and is responsible for a disproportionate number of reading disabilities and emotional traumas among children. The knowledge of phonetics is a century old at least. Means to achieve phonetic spelling are available in the two languages where the problem is most acute: any number of phonetic alphabets for English, and the much older, forty-eight-syllable Kana scripts in Japanese. For both countries there are examples next door of a successful shift to a phonetic script. The English have the successful model of German spelling reform of the mid-nineteenth century; the Japanese, the equally successful—and much earlier—phonetic reform of the Korean script. Yet in neither country is there the slightest receptivity for an innovation that, one would say, is badly needed, eminently rational, and proven by example to be safe, fairly easy, and efficacious. Why? Explanations abound, but no one really knows.

There is no way to eliminate the element of risk, no way even to reduce it. Market research does not work—one cannot do market research on something that does not exist. Opinion research is probably not just useless but likely to do damage. At least this is what the experience with "expert opinion" on the receptivity to knowledge-based innovation would indicate.

Yet there is no choice. If we want knowledge-based innovation, we must gamble on receptivity to it.

The risks are highest in innovations based on new knowledge in science and technology. They are particularly high, of course, in innovations in areas that are currently "hot"—personal computers, at the present time, or biotechnology. By contrast, areas that are not in the public eye have far lower risks, if only because there is more time. And in innovations where the knowledge base is not science or technology —social innovations, for instance—the risks are lower still. But high risk is inherent in knowledge-based innovation. It is the price we have to pay for its impact and above all for its capacity to bring about change, not only in products and services but in how we see the world, our place in it, and eventually ourselves.

Yet the risks even of high-tech innovation can be substantially reduced by integrating new knowledge as the source of innovation with one of the other sources defined earlier, the unexpected, incongruities, and especially process need. In these areas receptivity has either already been established or can be tested fairly easily and with good reliability. And in these areas, too, the knowledge or knowledges that have to be produced to complete an innovation can usually be defined with considerable precision. This is the reason why "program research" is becoming so popular. But even program research requires a great deal of system and self-discipline, and has to be organized and purposeful.

The demands on knowledge-based innovators are thus very great. They are also different from those in other areas of innovation. The risks they face are different, too; time, for instance, is not on their side. But if the risks are greater, so are the potential rewards. The other innovators may reap a fortune. The knowledge-based innovator can hope for fame as well.

10

The Bright Idea

Innovations based on a bright idea probably outnumber all other categories taken together. Seven or eight out of every ten patents belong here, for example. A very large proportion of the new businesses that are described in the books on entrepreneurs and entrepreneurships are built around "bright ideas": the zipper, the ballpoint pen, the aerosol spray can, the tab to open soft drink or beer cans, and many more. And what is called research in many businesses aims at finding and exploiting bright ideas, whether for a new flavor in breakfast cereals or soft drinks, for a better running shoe, or for yet one more nonscorching clothes iron.

Yet bright ideas are the riskiest and least successful source of innovative opportunities. The casualty rate is enormous. No more than one out of every hundred patents for an innovation of this kind earns enough to pay back development costs and patent fees. A far smaller proportion, perhaps as low as one in five hundred, makes any money above its out-of-pocket costs.

And no one knows which ideas for an innovation based on a bright idea have a chance to succeed and which ones are likely to fail. Why did the aerosol can succeed, for instance? And why did a dozen or more similar inventions for the uniform delivery of particles fail dismally? Why does one universal wrench sell and most of the many others disappear? Why did the zipper find acceptance and practically displace buttons, even though it tends to jam? (After all, a jammed zipper on a dress, jacket, or pair of trousers can be quite embarrassing.)

Attempts to improve the predictability of innovations based on bright ideas have not been particularly successful.

Equally unsuccessful have been attempts to identify the personal traits, behavior, or habits that make for a successful innovator. "Successful inventors," an old adage says, "keep on inventing. They play the odds. If they try often enough, they will succeed."

This belief that you'll win if only you keep on trying out bright ideas is, however, no more rational than the popular fallacy that to win the jackpot at Las Vegas one only has to keep on pulling the lever. Alas, the machine is rigged to have the house win 70 percent of the time. The more often you pull, the more often you lose.

There is actually no empirical evidence at all for the belief that persistence pays off in pursuing the "brilliant idea," just as there is no evidence of any "system" to beat the slot machines. Some successful inventors have had only one brilliant idea and then quit: the inventor of the zipper, for instance, or of the ballpoint pen. And there are hundreds of inventors around who have forty patents to their name, and not one winner. Innovators do, of course, improve with practice. But only if they practice the right method, that is, if they base their work on a systematic analysis of the sources of innovative opportunity.

The reasons for both the unpredictability and the high casualty rate are fairly obvious. Bright ideas are vague and elusive. I doubt that anyone except the inventor of the zipper ever thought that buttons or hooks-and-eyes were inadequate to fasten clothing, or that anyone but the inventor of the ballpoint pen could have defined what, if anything, was unsatisfactory about that nineteenth-century invention, the fountain pen. What need was satisfied by the electric toothbrush, one of the market successes of the 1960s? It still has to be hand-held, after all.

And even if the need can be defined, the solution cannot usually be specified. That people sitting in their cars in a traffic jam would like some diversion was perhaps not so difficult to figure out. But why did the small TV set which Sony developed around 1965 to satisfy this need fail in the marketplace, whereas the far more expensive car stereo succeeded? In retrospect, it is easy to answer this. But could it possibly have been answered in prospect?

The entrepreneur is therefore well advised to forgo innovations based on bright ideas, however enticing the success stories. After all, somebody wins a jackpot on the Las Vegas slot machines every week, yet the best any one slot-machine player can do is try not lose more than he or she can afford. Systematic, purposeful entrepreneurs analyze the systematic areas, the seven sources that I've discussed in Chapters 3 through 9.

There is enough in these areas to keep busy any one individual entrepreneur and any one entrepreneurial business or public-service institution. In fact, there is far more than anyone could possibly fully exploit. And in these areas we know how to look, what to look for, and what to do.

All one can do for innovators who go in for bright ideas is to tell them what to do should their innovation, against all odds, be successful. Then the rules for a new venture apply (see Chapter 15). And this is, of course, the reason why so much of the literature on entrepreneurship deals with starting and running the new venture rather than with innovation itself.

And yet an entrepreneurial economy cannot dismiss cavalierly the innovation based on a bright idea. The individual innovation of this kind is not predictable, cannot be organized, cannot be systematized, and fails in the overwhelming majority of cases. Also many, very many, are trivial from the start. There are always more patent applications for new can openers, for new wig stands, and for new belt buckles than for anything else. And in any list of new patents there is always at least one foot warmer than can double as a dish towel. Yet the volume of such bright-idea innovation is so large that the tiny percentage of successes represents a substantial source of new businesses, new jobs, and new performance capacity for the economy.

In the theory and practice of innovation and entrepreneurship, the bright-idea innovation belongs in the appendix. But it should be appreciated and rewarded. It represents qualities that society needs: initiative, ambition, and ingenuity. There is little society can do, perhaps, to promote such innovation. One cannot promote what one does not understand. But at least society should not discourage, penalize, or make difficult such innovations. Seen in this perspective, the recent trend in developed countries, and especially in the United States, to discourage the individual who tries to come up with a bright-idea innovation (by raising patent fees, for instance) and generally to discourage patents as "anticompetitive" is short-sighted and deleterious.

11

Principles of Innovation

I

All experienced physicians have seen "miracle cures." Patients suffering from terminal illnesses recover suddenly—sometimes spontaneously, sometimes by going to faith healers, by switching to some absurd diet, or by sleeping during the day and being up and about all night. Only a bigot denies that such cures happen and dismisses them as "unscientific." They are real enough. Yet no physician is going to put miracle cures into a textbook or into a course to be taught to medical students. They cannot be replicated, cannot be taught, cannot be learned. They are also extremely rare; the overwhelming majority of terminal cases do die, after all.

Similarly, there are innovations that do not proceed from the sources described in the preceding chapters, innovations that are not developed in any organized, purposeful, systematic manner. There are innovators who are "kissed by the Muses," and whose innovations are the result of a "flash of genius" rather than of hard, organized, purposeful work. But such innovations cannot be replicated. They cannot be taught and they cannot be learned. There is no known way to teach someone how to be a genius. But also, contrary to popular belief in the romance of invention and innovation, "flashes of genius" are uncommonly rare. What is worse, I know of not one such "flash of genius" that turned into an innovation. They all remained brilliant ideas.

The greatest inventive genius in recorded history was surely Leonardo da Vinci. There is a breathtaking idea—submarine or helicopter or automatic forge—on every single page of his notebooks. But not one of these could have been converted into an innovation with the technology and the materials of 1500. Indeed, for none of them would there

133

have been any receptivity in the society and economy of the time.

Every schoolboy knows of James Watt as the "inventor" of the steam engine, which he was not. Historians of technology know that Thomas Newcomen in 1712 built the first steam engine which actually performed useful work: it pumped the water out of an English coal mine. Both men were organized, systematic, purposeful innovators. Watt's steam engine in particular is the very model of an innovation in which newly available knowledge (how to ream a smooth cylinder) and the design of a "missing link" (the condenser) were combined into a process need-based innovation, the receptivity for which had been created by Newcomen's engine (several thousand were by then in use). But the true "inventor" of the combustion engine, and with it of what we call modern technology, was neither Watt nor Newcomen. It was the great Anglo-Irish chemist Robert Boyle, who did so in a "flash of genius." Only Boyle's engine did not work and could not have worked. For Boyle used the explosion of gunpower to drive the piston, and this so fouled the cylinder that it had to be taken apart and cleaned after each stroke. Boyle's idea enabled first Denis Papin (who had been Boyle's assistant in building the gunpowder engine), then Newcomen, and finally Watt, to develop a working combustion engine. All Boyle, the genius, had was a brilliant idea. It belongs in the history of ideas and not in the history of technology or of innovation.

The purposeful innovation resulting from analysis, system, and hard work is all that can be discussed and presented as the practice of innovation. But this is all that need be presented since it surely covers at least 90 percent of all effective innovations. And the extraordinary performer in innovation, as in every other area, will be effective only if grounded in the discipline and master of it.

What, then, are the principles of innovation, representing the hard core of the discipline? There are a number of "do's"—things that have to be done. There are also a few "dont's"—things that had better not be done. And then there are what I would call "conditions."

II

THE DO'S

1. Purposeful, systematic innovation begins with the analysis of the opportunities. It begins with thinking through what I have called the sources of innovative opportunities. In different areas, different sources

will have different importance at different times. Demographics, for instance, may be of very little concern to innovators in fundamental industrial processes, to someone looking, say, for the "missing link" in a process such as papermaking, where there is a clear incongruity between economic realities. New knowledge, by the same token, may be of very little relevance to someone innovating a new social instrument to satisfy a need created by changing demographics. But all the sources of innovative opportunity should be systematically analyzed and systematically studied. It is not enough to be alerted to them. The search has to be organized, and must be done on a regular, systematic basis.

2. Innovation is both conceptual and perceptual. The second imperative of innovation is therefore to go out to look, to ask, to listen. This cannot be stressed too often. Successful innovators use both the right side and the left side of their brains. They look at figures, and they look at people. They work out analytically what the innovation has to be to satisfy an opportunity. And then they go out and look at the customers, the users, to see what their expectations, their values, their needs are.

Receptivity can be perceived, as can values. One can perceive that this or that approach will not fit in with the expectations or the habits of the people who have to use it. And then one can ask: "What does this innovation have to reflect so that the people who have to use it will *want* to use it, and see in it *their* opportunity?" Otherwise one runs the risk of having the right innovation in the wrong form—as happened to the leading producer of computer programs for learning in American schools, whose excellent and effective programs were not used by teachers scared stiff of the computer, who perceived the machine as something that, far from being helpful, threatened them.

3. An innovation, to be effective, has to be simple and it has to be focused. It should do only one thing, otherwise, it confuses. If it is not simple, it won't work. Everything new runs into trouble; if complicated, it cannot be repaired or fixed. All effective innovations are breathtakingly simple. Indeed, the greatest praise an innovation can receive is for people to say: "This is obvious. Why didn't I think of it?"

Even the innovation that creates new uses and new markets should be directed toward a specific, clear, designed application. It should be focused on a specific need that it satisfies, on a specific end result that it produces.

4. Effective innovations start small. They are not grandiose. They try to do one specific thing. It may be to enable a moving vehicle to draw electric power while it runs along rails—the innovation that made possi-

ble the electric streetcar. Or it may be as elementary as putting the same number of matches into a matchbox (it used to be fifty), which made possible the automatic filling of matchboxes and gave the Swedish originators of the idea a world monopoly on matches for almost half a century. Grandiose ideas, plans that aim at "revolutionizing an industry," are unlikely to work.

Innovations had better be capable of being started small, requiring at first little money, few people, and only a small and limited market. Otherwise, there is not enough time to make the adjustments and changes that are almost always needed for an innovation to succeed. Initially innovations rarely are more than "almost right." The necessary changes can be made only if the scale is small and the requirements for people and money fairly modest.

5. But—and this is the final "do"—a successful innovation aims at leadership. It does not aim necessarily at becoming eventually a "big business"; in fact, no one can foretell whether a given innovation will end up as a big business or a modest achievement. But if an innovation does not aim at leadership from the beginning, it is unlikely to be innovative enough, and therefore unlikely to be capable of establishing itself. Strategies (to be discussed in Chapters 16 through 19) vary greatly, from those that aim at dominance in an industry or a market to those that aim at finding and occupying a small "ecological niche" in a process or market. But all entrepreneurial strategies, that is, all strategies aimed at exploiting an innovation, must achieve leadership within a given environment. Otherwise they will simply create an opportunity for the competition.

III

THE DONT'S

And now the few important "dont's."

1. The first is simply not to try to be clever. Innovations have to be handled by ordinary human beings, and if they are to attain any size and importance at all, by morons or near-morons. Incompetence, after all, is the only thing in abundant and never-failing supply. Anything too clever, whether in design or execution, is almost bound to fail.

2. Don't diversify, don't splinter, don't try to do too many things at once. This is, of course, the corollary to the "do": be focused! Innova-

tions that stray from a core are likely to become diffuse. They remain ideas and do not become innovations. The core does not have to be technology or knowledge. In fact, market knowledge supplies a better core of unity in any enterprise, whether business or public-service institution, than knowledge or technology do. But there has to be a core of unity to innovative efforts or they are likely to fly apart. An innovation needs the concentrated energy of a unified effort behind it. It also requires that the people who put it into effect understand each other, and this, too, requires a unity, a common core. This, too, is imperiled by diversity and splintering.

3. Finally, don't try to innovate for the future. Innovate for the present! An innovation may have long-range impact; it may not reach its full maturity until twenty years later. The computer, as we have seen, did not really begin to have any sizable impact on the way business was being done until the early 1970s, twenty-five years after the first working models were introduced. But from the first day the computer had some specific current applications, whether scientific calculation, making payroll, or simulation to train pilots to fly airplanes. It is not good enough to be able to say, "In twenty-five years there will be so many very old people that they will need this." One has to be able to say, "There are enough old people around today for this to make a difference to them. Of course, time is with us—in twenty-five years there will be many more." But unless there is an immediate application in the present, an innovation is like the drawings in Leonardo da Vinci's notebook—a "brilliant idea." Very few of us have Leonardo's genius and can expect that our notebooks alone will assure immortality.

The first innovator who fully understood this third caveat was probably Edison. Every other electrical inventor of the time began to work around 1860 or 1865 on what eventually became the light bulb. Edison waited for ten years until the knowledge became available; up to that point, work on the light bulb was "of the future." But when the knowledge became available—when, in other words, a light bulb could become "the present"—Edison organized his tremendous energies and an extraordinarily capable staff and concentrated for a couple of years on that one innovative opportunity.

Innovative opportunities sometimes have long lead times. In pharmaceutical research, ten years of research and development work are by no means uncommon or particularly long. And yet no pharmaceutical company would dream of starting a research project for something which does not, if successful, have immediate application as a drug for health-care needs that already exist.

THREE CONDITIONS

Finally, there are three conditions. All three are obvious but often go disregarded.

1. *Innovation is work.* It requires knowledge. It often requires great ingenuity. There are clearly people who are more talented innovators than the rest of us. Also, innovators rarely work in more than one area. For all his tremendous innovative capacity, Edison worked only in the electrical field. And an innovator in financial areas, Citibank in New York, for instance, is unlikely to embark on innovations in retailing or health care. In innovation as in any other work there is talent, there is ingenuity, there is predisposition. But when all is said and done, innovation becomes hard, focused, purposeful work making very great demands on diligence, on persistence, and on commitment. If these are lacking, no amount of talent, ingenuity, or knowledge will avail.

2. *To succeed, innovators must build on their strengths.* Successful innovators look at opportunities over a wide range. But then they ask, "Which of these opportunities fits *me*, fits *this* company, puts to work what we (or I) are good at and have shown capacity for in performance?" In this respect, of course, innovation is no different from other work. But it may be more important in innovation to build on one's strengths because of the risks of innovation and the resulting premium on knowledge and performance capacity. And in innovation, as in any other venture, there must also be a temperamental "fit." Businesses do not do well in something they do not really respect. No pharmaceutical company—run as it has to be by scientifically minded people who see themselves as "serious"—has done well in anything so "frivolous" as lipsticks or perfumes. Innovators similarly need to be temperamentally attuned to the innovative opportunity. It must be important to them and make sense to them. Otherwise they will not be willing to put in the persistent, hard, frustrating work that successful innovation always requires.

3. And finally, *innovation is an effect in economy and society,* a change in the behavior of customers, of teachers, of farmers, of eye surgeons—of people in general. Or it is a change in a process—that is,

in how people work and produce something. Innovation therefore always has to be close to the market, focused on the market, indeed market-driven.

THE CONSERVATIVE INNOVATOR

A year or two ago I attended a university symposium on entrepreneurship at which a number of psychologists spoke. Although their papers disagreed on everything else, they all talked of an "entrepreneurial personality," which was characterized by a "propensity for risk-taking."

A well-known and successful innovator and entrepreneur who had built a process-based innovation into a substantial worldwide business in the space of twenty-five years was then asked to comment. He said: "I find myself baffled by your papers. I think I know as many successful innovators and entrepreneurs as anyone, beginning with myself. I have never come across an 'entrepreneurial personality.' The successful ones I know all have, however, one thing—and only one thing—in common: they are *not* 'risk-takers.' They try to define the risks they have to take and to minimize them as much as possible. Otherwise none of us could have succeeded. As for myself, if I had wanted to be a risk-taker, I would have gone into real estate or commodity trading, or I would have become the professional painter my mother wanted me to be."

This jibes with my own experience. I, too, know a good many successful innovators and entrepreneurs. Not one of them has a "propensity for risk-taking."

The popular picture of innovators—half pop-psychology, half Hollywood—makes them look like a cross between Superman and the Knights of the Round Table. Alas, most of them in real life are unromantic figures, and much more likely to spend hours on a cash-flow projection than to dash off looking for "risks." Of course innovation is risky. But so is stepping into the car to drive to the supermarket for a loaf of bread. All economic activity is by definition "high-risk." And defending yesterday—that is, not innovating—is far more risky than making tomorrow. The innovators I know are successful to the extent to which they define risks and confine them. They are successful to the extent to which they systematically analyze the sources of innovative opportunity, then pinpoint the opportunity and exploit it. Whether opportunities of small and clearly definable risk, such as exploiting the unexpected or a process need, or opportunities of much greater but still definable risk, as in knowledge-based innovation.

Successful innovators are conservative. They have to be. They are not "risk-focused"; they are "opportunity-focused."

II

THE PRACTICE OF
ENTREPRENEURSHIP

———

The entrepreneurial requires different management from the existing. But like the existing it requires systematic, organized, purposeful management. And while the ground rules are the same for every entrepreneurial organization, the existing business, the public-service institution, and the new venture present different challenges, have different problems, and have to guard against different degenerative tendencies. There is need also for individual entrepreneurs to face up to decisions regarding their own roles and their own commitments.

12

Entrepreneurial Management

Entrepreneurship is based on the same principles, whether the entrepreneur is an existing large institution or an individual starting his or her new venture singlehanded. It makes little or no difference whether the entrepreneur is a business or a nonbusiness public-service organization, nor even whether the entrepreneur is a governmental or nongovernmental institution. The rules are pretty much the same, the things that work and those that don't are pretty much the same, and so are the kinds of innovation and where to look for them. In every case there is a discipline we might call *Entrepreneurial Management.*

Yet the existing business faces different problems, limitations, and constraints from the solo entrepreneur, and it needs to learn different things. The existing business, to oversimplify, knows how to manage but needs to learn how to be an entrepreneur and how to innovate. The nonbusiness public-service institution, too, faces different problems, has different learning needs, and is prone to making different mistakes. And the new venture needs to learn how to be an entrepreneur and how to innovate, but above all, it needs to learn how to manage.

For each of these three:

- the existing business
- the public-service institution
- the new venture

a specific guide to the practice of entrepreneurship must be developed. What does each have to do? What does each have to watch for? And what had each better avoid doing?

Logically, the discussion might start with the new venture, just as, logically, the study of medicine might start with the embryo and newborn baby. But the medical student starts out by studying the anatomy and pathology of the adult, and the practice of entrepreneurship is likewise best started by discussing the "adult," the existing business and the policies, practices and problems that are pertinent in managing it for entrepreneurship.

Today's businesses, especially the large ones, simply will not survive in this period of rapid change and innovation unless they acquire entrepreneurial competence. In this respect the late twentieth century is totally different from the last great entrepreneurial period in economic history, the fifty or sixty years that came to an end with the outbreak of World War I. There were not many big businesses around in those years, and not even many middle-sized ones. Today, it is not only in the self-interest of the many existing big businesses to learn to manage themselves for entrepreneurship; they have a social responsibility to do so. In sharp contrast to the situation a century ago, rapid destruction of the existing businesses—especially the big ones—by innovation, the "creative destruction" by the innovator, in Joseph Schumpeter's famous phrase, poses a genuine social threat today to employment, to financial stability, to social order, and to governmental responsibility.

Existing businesses will need to change, and change greatly in any event. Within twenty-five years (see Chapter 7) every industrially developed non-Communist country will see the blue-collar labor force engaged in manufacturing shrink to one-third of what it is now, while manufacturing output should go up three- or four-fold—a development that will parallel the development in agriculture in the industrialized non-Communist countries during the twenty-five years following World War II. In order to impart stability and leadership in a transition of this magnitude, existing businesses will have to learn how to survive, indeed, how to propser. And that they can only do if they learn to be successful entrepreneurs.

In many cases, the entrepreneurship needed can only come from existing businesses. Some of the giants of today may well not survive the next twenty-five years. But we now know that the medium-sized business is particularly well positioned to be a successful entrepreneur and innovator, provided only that it organize itself for entrepreneurial management. It is the existing business—and the fair-sized rather than the small one—that has the best capability for entrepreneurial leadership. It has the necessary resources, especially the human resources. It has already acquired managerial competence and built a management team. It has both the opportunity and the responsibility for effective entrepreneurial management.

The same holds true for the public-service institutions, and especially for those discharging nonpolitical functions, whether owned by government and financed by tax money or not; for hospitals, schools, and universities; for the public services of local governments; for community agencies and volunteer organizations such as the Red Cross, the Boy Scouts, and the Girl Scouts; for churches and church-related organizations; but also for professional and trade associations, and many more. A period of rapid change makes obsolete a good many of the old concerns, or at least makes ineffectual a good many of the ways in which they have been addressed. At the same time, such a period creates opportunities for tackling new tasks, for experimentation, and for social innovation.

Above all, there has been a major change in perception and mood in the public domain (cf. Chapter 8). A hundred years ago, the "panic" of 1873 brought to an end the century of *laissez faire* that had begun with Adam Smith's *Wealth of Nations* in 1776. For a hundred years from 1873 on, being "modern," "progressive," or "forward-looking" meant looking to government as the agent of social change and betterment. For better or worse, that period has come to an end in all non-Communist developed countries (and probably in the developed Communist countries as well). We do not yet know what the next wave of "progressivism" will be. But we do know that anyone who still preaches the "liberal" or "progressive" gospel of 1930—or even of 1960, of the Kennedy and Johnson years—is not a "progressive" but a "reactionary." We do not know whether privatization,* that is, turning activities back from government to nongovernmental operation (albeit not necessarily to operation by a business enterprise, as most people have interpreted the term) will work or will go very far. But we do know that no non-Communist developed country will move further toward nationalization and governmental control out of hope, expectation, and belief in the traditional promises. It will do so only out of frustration and with a sense of failure. And this is a situation in which public-service institutions have both an opportunity and a responsibility to be entrepreneurial and to innovate.

But precisely because they are public-service institutions, they face specific different obstacles and challenges, and are prone to making

---

*A word that I coined in 1969 in *The Age of Discontinuity* (New York: Harper & Row; London: William Heinemann).

different mistakes. Entrepreneurship in the public-service institution thus needs to be discussed separately.

Finally, there is the new venture. This will continue to be a main vehicle for innovation, as it has been in all major entrepreneurial periods and is again today in the new entrepreneurial economy of the United States. There is indeed no lack of would-be entrepreneurs in the United States, no shortage of new ventures. But most of them, especially the high-tech ones, have a great deal to learn about entrepreneurial management and will have to learn it if they are to survive.

The gap between the performance of the average practitioner and that of the leaders in entrepreneurship and innovation is enormous in all three categories. Fortunately, there are enough examples around of the successful practice of entrepreneurship to make possible a systematic presentation of entrepreneurial management that is both practice and theory, both description and prescription.

13

The Entrepreneurial Business

"Big businesses don't innovate," says the conventional wisdom. This sounds plausible enough. True, the new, major innovations of this century did not come out of the old, large businesses of their time. The railroads did not spawn the automobile or the truck; they did not even try. And though the automobile companies did try (Ford and General Motors both pioneered in aviation and aerospace), all of today's large aircraft and aviation companies have evolved out of separate new ventures. Similarly, today's giants of the pharmaceutical industry are, in the main, companies that were small or nonexistent fifty years ago when the first modern drugs were developed. Every one of the giants of the electrical industry—General Electric, Westinghouse, and RCA in the United States; Siemens and Philips on the Continent; Toshiba in Japan —rushed into computers in the 1950s. Not one was successful. The field is dominated by IBM, a company that was barely middle-sized and most definitely not high-tech forty years ago.

And yet the all but universal belief that large businesses do not and cannot innovate is not even a half-truth; rather, it is a misunderstanding.

In the first place, there are plenty of exceptions, plenty of large companies that have done well as entrepreneurs and innovators. In the United States, there is Johnson & Johnson in hygiene and health care, and 3M in highly engineered products for both industrial and consumer markets. Citibank, America's and the world's largest non-governmental financial institution, well over a century old, has been a major innovator in many areas of banking and finance. In Germany, Hoechst—one of the world's largest chemical companies, and more than 125 years old by now—has become a successful innovator in the

147

pharmaceutical industry. In Sweden, ASEA, founded in 1884 and for the last sixty or seventy years a very big company, is a true innovator in both long-distance transmission of electrical power and robotics for factory automation.

To confuse things even more there are quite a few big, older businesses that have succeeded as entrepreneurs and innovators in some fields while failing dismally in others. The (American) General Electric Company failed in computers, but has been a successful innovator in three totally different fields: aircraft engines, engineered inorganic plastics, and medical electronics. RCA also failed in computers but succeeded in color television. Surely things are not quite as simple as the conventional wisdom has it.

Secondly, it is not true that "bigness" is an obstacle to entrepreneurship and innovation. In discussions of entrepreneurship one hears a great deal about the "bureaucracy" of big organizations and of their "conservatism." Both exist, of course, and they are serious impediments to entrepreneurship and innovation—but to all other performance just as much. And yet the record shows unambiguously that among existing enterprises, whether business or public-sector institutions, the small ones are least entrepreneurial and least innovative. Among existing entrepreneurial businesses there are a great many very big ones; the list above could have been enlarged without difficulty to one hundred companies from all over the world, and a list of innovative public-service institutions would also include a good many large ones.

And perhaps the most entrepreneurial business of them all is the large middle-sized one, such as the American company with $500 million in sales in the mid-1980s.* But *small* existing enterprises would be conspicuously absent from any list of entrepreneurial businesses.

It is not size that is an impediment to entrepreneurship and innovation; it is the existing operation itself, and especially the existing *successful* operation. And it is easier for a big or at least a fair-sized company to surmount this obstacle than it is for a small one. Operating anything—a manufacturing plant, a technology, a product line, a distribution system—requires constant effort and unremitting atten-

---

*This has long been suspected. Now, however, conclusive evidence is available in the study of one hundred medium-sized "growth" companies by Richard E. Cavenaugh and Donald K. Clifford, Jr., "Lessons from America's Mid-Sized Growth Companies," *McKinsey Quarterly* (Autumn 1983).

tion. The one thing that can be guaranteed in any kind of operation is the daily crisis. The daily crisis cannot be postponed, it has to be dealt with right away. And the existing operation demands high priority and deserves it.

The new always looks so small, so puny, so unpromising next to the size and performance of maturity. Anything truly new that looks big is indeed to be distrusted. The odds are heavily against its succeeding. And yet successful innovators, as was argued earlier, start small and, above all, simple.

The claim of so many businesses, "Ten years from now, ninety percent of our revenues will come from products that do not even exist today," is largely boasting. Modifications of existing products, yes; variations, yes; even extensions of existing products into new markets and new end uses—with or without modifications. But the truly new venture tends to have a longer lead time. Successful businesses, businesses that are today in the right markets with the right products or services, are likely ten years hence to get three-quarters of their revenues from products and services that exist today, or from their linear descendants. In fact, if today's products or services do not generate a continuing and large revenue stream, the enterprise will not be able to make the substantial investment in tomorrow that innovation requires.

It thus takes special effort for the existing business to become entrepreneurial and innovative. The "normal" reaction is to allocate productive resources to the existing business, to the daily crisis, and to getting a little more out of what we already have. The temptation in the existing business is always to feed yesterday and to starve tomorrow.

It is, of course, a deadly temptation. The enterprise that does not innovate inevitably ages and declines. And in a period of rapid change such as the present, an entrepreneurial period, the decline will be fast. Once an enterprise or an industry has started to look back, turning it around is exceedingly difficult, if it can be done at all. But the obstacle to entrepreneurship and innovation which the success of the present business constitutes is a real one. The problem is precisely that the enterprise is so successful, that it is "healthy" rather than degeneratively diseased by bureaucracy, red tape, or complacency.

This is what makes the examples of existing businesses that do manage successfully to innovate so important, and especially the examples of existing large and fair-sized businesses that are also successful entrepreneurs and innovators. These businesses show that the obstacle of success, the obstacle of the existing, *can* be overcome. And it can be overcome in such a way that both the existing and the new, the mature and the infant, benefit and prosper. The large companies that are successful entrepreneurs and innovators—Johnson & Johnson, Hoechst, ASEA, 3M, or the one hundred middle-sized "growth" companies—clearly know how to do it.

Where the conventional wisdom goes wrong is in its assumption that entrepreneurship and innovation are natural, creative, or spontaneous. If entrepreneurship and innovation do not well up in an organization, something must be stifling them. That only a minority of existing successful businesses are entrepreneurial and innovative is thus seen as conclusive evidence that existing businesses quench the entrepreneurial spirit.

But entrepreneurship is not "natural"; it is not "creative." It is work. Hence, the correct conclusion from the evidence is the opposite of the one commonly reached. That a substantial number of existing businesses, and among them a goodly number of fair-sized, big, and very big ones, succeed as entrepreneurs and innovators indicates that entrepreneurship and innovation can be achieved by any business. But they must be consciously striven for. They can be learned, but it requires effort. Entrepreneurial businesses treat entrepreneurship as a duty. They are disciplined about it . . . they work at it . . . they practice it.

Specifically, entrepreneurial management requires *policies and practices* in four major areas.

First, the organization must be made receptive to innovation and willing to perceive change as an opportunity rather than a threat. It must be organized to do the hard work of the entrepreneur. Policies and practices are needed to create the entrepreneurial climate.

Second, systematic measurement or at least appraisal of a company's performance as entrepreneur and innovator is mandatory, as well as built-in learning to improve performance.

Third, entrepreneurial management requires specific practices pertaining to organizational structure, to staffing and managing, and to compensation, incentives, and rewards.

Fourth, there are some "dont's": things *not to do* in entrepreneurial management.

II

A Latin poet called the human being *"rerum novarum cupidus* (greedy for new things)."* Entrepreneurial management must make each manager of the existing business *"rerum novarum cupidus."*

"How can we overcome the resistance to innovation in the existing organization?" is a question commonly asked by executives. Even if we knew the answer, it would still be the wrong question. The right one is: "How can we make the organization receptive to innovation, want innovation, reach for it, work for it?" When innovation is perceived by the organization as something that goes against the grain, as swimming against the current, if not as a heroic achievement, there will be no innovation. Innovation must be part and parcel of the ordinary, the norm, if not routine.

This requires specific policies. First, innovation, rather than holding on to what already exists, must be made attractive and beneficial to managers. There must be clear understanding throughout the organization that innovation is the best means to preserve and perpetuate that organization, and that it is the foundation for the individual manager's job security and success.

Second, the importance of the need for innovation and the dimensions of its time frame must be both defined and spelled out.

And finally, there needs to be an innovation plan, with specific objectives laid out.

1. There is only one way to make innovation attractive to managers: a systematic policy of abandoning whatever is outworn, obsolete, no longer productive, as well as the mistakes, failures, and misdirections of effort. Every three years or so, the enterprise must put every single product, process, technology, market, distributive channel, not to mention every single internal staff activity, on trial for its life. It must ask: Would we *now* go into this product, this market, this distributive channel, this technology *today?* If the answer is "No," one does not respond with, "Let's make another study." One asks, "What do we have to do to stop wasting resources on this product, this market, this distributive channel, this staff activity?"

Sometimes abandonment is not the answer, and may not even be possible. But then at least one limits further efforts and makes sure that productive resources of men and money are no longer devoured by yesterday. This is the right thing to do in any event to maintain the health of the organization: every organism needs to eliminate its waste products or else it poisons itself. It is, however, an absolute necessity, if an enterprise is to be capable of innovation and is to be receptive to it. "Nothing so powerfully concentrates a man's mind as to know that he will be hung on the morning," Dr. Johnson was fond of saying. Nothing so powerfully concentrates a manager's mind on innovation as the knowledge that the present product or service will be abandoned within the foreseeable future.

Innovation requires major effort. It requires hard work on the part of performing, capable people—the scarcest resource in any organization. "Nothing requires more heroic efforts than to keep a corpse from stinking, and yet nothing is quite so futile," is an old medical proverb. In almost any organization I have come across, the best people are engaged in this futile effort; yet all they can hope to accomplish is to delay acceptance of the inevitable a little longer and at great cost.

But if it is known throughout the organization that the dead will be left to bury their dead, then the living will be willing—indeed, eager —to go to work on innovation.

To allow it to innovate, a business has to be able to free its best performers for the challenges of innovation. Equally it has to be able to devote financial resources to innovation. It will not be able to do either unless it organizes itself to slough off alike the successes of the past, the failures, and especially the "near-misses," the things that "should have worked" but didn't. If executives know that it is company policy to abandon, then they will be motivated to look for the new, to encourage entrepreneurship, and will accept the need to become entrepreneurial themselves. This is the first step—a form of organizational hygiene.

2. The second step, the second policy needed to make an existing business "greedy for new things," is to face up to the fact that all existing products, services, markets, distributive channels, processes, technologies, have limited—and usually short—health and life expectancies.

An analysis of the life cycle of existing products, services, and so on has become popular since the 1970s. Some examples are the strategy concepts advocated by the Boston Consulting group; the books on strategy by the Harvard Business School professor Michael Porter; and so-called portfolio management.*

In the strategies that have been widely advertised these last ten years, especially portfolio management, the findings of such analysis constitute an action program by themselves. This is a misunderstanding and bound to lead to disappointing results, as a good many companies found out when they rushed into such strategies in the late 1970s and early 1980s. The findings should lead to a *diagnosis*. This in turn requires judgment. It requires knowledge of the business, of its products, its markets, its customers, its technologies. It requires experience rather than analysis alone. The idea that bright young people straight from business school and equipped only with sharp analytical tools could crunch out of their computer life-and-death decisions about businesses, products, and markets is pure quackery, to be blunt.

This analysis (in *Managing for Results*, I called it a "Business X-Ray") is intended as a tool to find the right questions rather than a way automatically to come up with the right answers. It is a challenge to all the knowledge that can be found in a given company, and all the experience. It will—and should—provoke dissent. The action that follows from classifying this or that product as "today's breadwinner" is a *risk-taking decision*. And so is what to do with the product that is on the point of becoming "yesterday's breadwinner," or with an "unjustified specialty," or with an "investment in managerial ego."†

3. The Business X-Ray furnishes the information needed to define how much innovation a given business requires, in what areas, and within what time frame. The best and simplest approach to this was developed by Michael J. Kami as a member of the Entrepreneurship Seminar at the New York University Graduate Business School in the 1950s. Kami first applied his approach to IBM, where he served as head of business planning; and then, in the early 1960s, to Xerox, where he served for several years in a similar capacity.

In this approach a company lists each of its products or services, but

---

*All these approaches have their origin in a book of mine published twenty years ago, *Managing for Results* (New York: Harper & Row, 1964), the first systematic work on business strategy, to my knowledge. This in turn grew out of the Entrepreneurship Seminar I ran in the late fifties at New York University. The analysis presented in *Managing for Results* (Chapters 1–5), with its ranking of all products and services into a small number of categories according to their performance, characteristics, and life expectancies, is still a useful tool for the analysis of product-life and product-health.

†For a definition of these terms, see *Managing for Results*, especially Chapter 4, How Are We Doing?, pp. 51–68.

also the markets each serves and the distributive channels it uses, in order to estimate their position on the product life cycle. How much longer will this product still grow? How much longer will it still maintain itself in the marketplace? How soon can it be expected to age and decline—and how fast? When will it become obsolescent? This enables the company to estimate where it would be if it confined itself to managing to the best of its ability what already exists. And this then shows the gap between what can be expected realistically, and what a company still needs to do to achieve its objectives, whether in sales, in market standing, or in profitability.

The gap is the minimum that must be filled if the company is not to go downhill. In fact, the gap has to be filled or the company will soon start to die. The entrepreneurial achievement must be large enough to fill the gap, and timely enough to fill it before the old becomes obsolescent.

But innovative efforts do not carry certainty; they have a high probability of failure and an even higher one of delay. A company therefore should have under way at least three times the innovative efforts which, if successful, would fill the gap.

Most executives consider this excessively high. Yet experience has proved that it errs on the low side, if it errs at all. To be sure, some innovative efforts will do better than anyone expects, but others will do much less well. And everything takes longer than we hope or estimate; everything also requires more effort. Finally, the one thing certain about any major innovative effort is that there are going to be last-minute hitches and last-minute delays. To demand innovative efforts which, if everything goes according to plan, yield three times the minimum results needed is only elementary precaution.

4. Systematic abandonment; the Business X-Ray of the existing business, its products, its services, its markets, its technologies; and the definition of innovation gap and innovation need—these together enable a company to formulate an *entrepreneurial plan* with objectives for innovation and deadlines.

Such a plan ensures that the innovation budget is adequate. And—the most important result of all—it determines how many people are needed, with what abilities and capacities. Only when people with proven performance capacity have been assigned to a project, supplied with the tools, the money, and the information they need to do the work, and given clear and unambiguous deadlines—only then do we have a plan. Until then, we have "good intentions," and what those are good for, everybody knows.

These are the fundamental policies needed to endow a business with entrepreneurial management; to make a business and its management greedy for new things; to make it perceive innovation as the healthy, normal, necessary course of action. Because it is based on a "Business X-Ray"—that is, on an analysis and diagnosis of the current business, its products, services, and markets—this approach also ensures that the existing business will not be neglected in the search for the new, and that the opportunities inherent in the existing products, services, and markets will not be sacrificed to the fascination with novelty.

The Business X-Ray is a tool for decision making. It enables us, indeed forces us, to allocate resources to results in the existing business. But it also makes it possible for us to determine how much is needed to create the business of tomorrow and its new products, new services, and new markets. It enables us to turn innovative intentions into innovative performance.

To render an existing business entrepreneurial, management must take the lead in making obsolete its own products and services rather than waiting for a competitor to do so. The business must be managed so as to perceive in the new an opportunity rather than a threat. It must be managed to work *today* on the products, services, processes, and technologies that will make a different tomorrow.

III

ENTREPRENEURIAL PRACTICES

Entrepreneurship in the existing business also requires managerial practices.

1. First among these, and the simplest, is focusing managerial vision on opportunity. People see what is presented to them; what is not presented tends to be overlooked. And what is presented to most managers are "problems"—especially in the areas where performance falls below expectations—which means that managers tend not to see the opportunities. They are simply not being presented with them.

Management, even in small companies, usually get a report on operating performance once a month. The first page of this report always lists the areas in which performance has fallen below budget, in which there is a "shortfall," in which there is a "problem." At the monthly management meeting, everyone then goes to work on the so-called problems. By the time the meeting adjourns for lunch, the whole morning has been taken up with the discussion of those problems.

Of course, problems have to be paid attention to, taken seriously, and tackled. But if they are the only thing that is being discussed, opportunities will die of neglect. In businesses that want to create receptivity to entrepreneurship, special care is therefore taken that the opportunities are also attended to (cf. Chapter 3 on the unexpected success).

In these companies, the operating report has *two* "first pages": the traditional one lists the problems; the other one lists all the areas in which performance is better than expected, budgeted, or planned for. For, as was stressed earlier, the unexpected success in one's own business is an important symptom of innovative opportunity. If it is not seen as such, the business is altogether unlikely to be entrepreneurial. In fact the business and its managers, in focusing on the "problems," are likely to brush aside the unexpected success as an intrusion on their time and attention. They will say, "Why should we do anything about it? It's going well without our messing around with it." But this only creates an opening for the competitor who is a little more alert and a little less arrogant.

Typically, in companies that are managed for entrepreneurship, there are therefore two meetings on operating results: one to focus on the problems and one to focus on the opportunities.

One medium-sized supplier of health-care products to physicians and hospitals, a company that has gained leadership in a number of new and promising fields, holds an "operations meeting" the second and the last Monday of each month. The first meeting is devoted to problems —to all the things which, in the last month, have done less well than expected or are still doing less well than expected six months later. This meeting does not differ one whit from any other operating meeting. But the second meeting—the one on the last Monday—discusses the areas where the company is doing better than expected: the sales of a given product that have grown faster than projected, or the orders for a new product that are coming in from markets for which it was not designed. The top management of the company (which has grown ten-fold in twenty years) believes that its success is primarily the result of building this opportunity focus into its monthly management meetings. "The opportunities we spot in there," the chief executive officer has said many times, "are not nearly as important as the entrepreneurial attitude which the habit of looking for opportunities creates throughout the entire management group."

2. This company follows a second practice to generate an entrepreneurial spirit throughout its entire management group. Every six months it holds a two-day management meeting for all executives in charge of divisions, markets, and major product lines—a group of about forty or fifty people. The first morning is set aside for reports to the entire group from three or four executives whose units have done exceptionally well as entrepreneurs and innovators during the past year. They are expected to report on what explains their success: "What did we do that turned out to be successful?" "How did we find the opportunity?" "What have we learned, and what entrepreneurial and innovative plans do we have in hand now?"

Again, what actually is reported in these sessions is less important than the impact on attitudes and values. But the operating managers in the company also stress how much they learn in each of these sessions, how many new ideas they get, and how they return back home from these sessions full of plans and eager to try them.

Entrepreneurial companies always look for the people and units that do better and do differently. They single them out, feature them, and constantly ask them: "What are you doing that explains your success?" "What are you doing that the rest of us aren't doing, and what are you *not* doing that the rest of us are?"

3. A third practice, and one that is particularly important in the large company, is a session—informal but scheduled and well prepared—in which a member of the top management group sits down with the junior people from research, engineering, manufacturing, marketing, accounting and so on. The senior opens the session by saying: "I'm not here to make a speech or to tell you anything, I'm here to listen. I want to hear from you what your aspirations are, but above all, where you see opportunities for this company and where you see threats. And what are your ideas for us to try to do new things, develop new products, design new ways of reaching the market? What questions do you have about the company, its policies, its direction . . . its position in the industry, in technology, in the marketplace?"

These sessions should not be held too often; they are a substantial time-burden on senior people. No senior executive should therefore be expected to sit down more than three times a year for a long afternoon or evening with a group of perhaps twenty-five or thirty juniors. But the sessions should be maintained systematically. They are an excellent vehicle for upward communications, the best means to enable juniors, and especially professionals, to look up from their narrow specialties and see the whole enterprise. They enable juniors to understand what top management is concerned with, and why. In turn, they give the seniors badly needed insight into the values, vision, and concerns of their younger colleagues. Above all, these sessions are one of the most effective ways to instill entrepreneurial vision throughout the company.

This practice has one built-in requirement. Those who suggest anything new, or even a change in the way things are being done, whether in respect to product or process, to market or service, should be expected to *go to work*. They should be asked to submit, within a reasonable period, a working paper to the presiding senior and to their colleagues in the session, in which they try to develop their idea. What would it look like if converted into reality? What in turn does reality have to look like for the idea to make sense? What are the assumptions regarding customers and markets, and so on. How much work is needed . . . how much money and how many people . . . and how much time? And what results might be expected?

Again, the yield of entrepreneurial ideas from all this may not be its most important product—though in many organizations the yield has been consistently high. The most valuable achievement may well be entrepreneurial vision, receptivity to innovation, and "greed for new things" throughout the entire organization.

IV

MEASURING INNOVATIVE PERFORMANCE

For a business to be receptive to entrepreneurship, innovative performance must be included among the measures by which that business controls itself. Only if we assess the entrepreneurial performance of a business will entrepreneurship become action. Human beings tend to behave as they are expected to.

In the normal assessments of a business, innovative performance is conspicuous by its absence. Yet it is not particularly difficult to build measurement, or at least judgment, of entrepreneurial and innovative performance into the controls of the business.

1. The first step builds into each innovative project feedback from results to expectations. This indicates the quality and reliability of both our innovative plans and our innovative efforts.

Research managers long ago learned to ask at the beginning of any research project: "What results do we expect from this project? When do we expect those results? When do we appraise the progress of the project so that we have control?" They have also learned to check whether their expectations are borne out by the actual course of events. This shows them whether they are tending to be too optimistic or too pessimistic, whether they expect results too soon or are willing to wait too long, whether they are inclined either to overestimate the impact of a successfully concluded research project or to underestimate it. And this in turn enables them to correct said tendencies, and to identify both the areas in which they do well and the ones in which they tend to do poorly. Such feedback is, of course, needed for all innovative efforts, not merely for technical research and development.

The first aim is to find out what we are doing well, for one can always go ahead and do more of the same, even if we usually do not have the slightest idea why we are doing well in a given area. Next, one finds out the limitations on one's strengths: for instance, a tendency either to underestimate the amount of time needed or to overestimate it; or a tendency to overestimate the amount of research required in a given area while underestimating the resources required for developing the results of research into a product or a process. Or one finds a tendency, very common and very damaging, to slow down marketing or promotion efforts for the new venture just when it is about to take off.

One of the most successful of the world's major banks attributes its achievements to the feedback it builds into all new efforts, whether it is going into a new market such as South Korea, into equipment leasing, or into issuing credit cards. By building feedback from results to expectations for all new endeavors, the bank and its top management have also learned what they can expect from new ventures: How soon a new effort can be expected to produce results and when it should be supported by greater efforts and greater resources.

Such feedback is needed for all innovative efforts, the development and introduction of a new safety program, say, or a new compensation plan. What are the first indications that the new effort is likely to get into trouble and needs to be reconsidered? And what are the indications that enable us to say that this effort, even though it looks as if it were headed for trouble, is actually doing all right, but also that it may take more time than we originally anticipated?

2. The next step is to develop a systematic review of innovative efforts all together. Every few years an entrepreneurial management looks at all the innovative efforts of the business. Which ones should receive more support at this stage and should be pushed? Which ones have opened up new opportunities? Which ones, on the other hand, are not doing what we expected them to do, and what action should we take? Has the time come to abandon them, or, on the contrary, has the time come to redouble our efforts—but with what expectations and what deadline?

The top management people at one of the world's largest and most successful pharmaceutical companies sit down once a year to review its innovative efforts. First, they review every new drug development, asking: "Is this development going in the right direction and at the right speed? Is it leading to something we want to put into our own line, or is it going to be something that won't fit our markets so we'd better license it to another pharmaceutical manufacturer? Or ought we perhaps abandon it?" And then the same people look at all the other innovative efforts, especially in marketing, asking exactly the same questions. Finally, they review, equally carefully, the innovative performance of their major competitors. In terms of its research budget and its total expenditures for innovation, this company ranks only in the middle level. Its record as an innovator and entrepreneur is, however, outstanding.

3. Finally, entrepreneurial management entails judging the company's total innovative performance against the company's innovative objectives, against its performance and standing in the market, and against its performance as a business all together.

Every five years, perhaps, top management sits down with its associates in each major area and asks: "What have you contributed to this company in the past five years that really made a difference? And what do you plan to contribute in the next five years?"

But are not innovative efforts by their nature intangible? How can one measure them?

It is indeed true that there are some areas in which no one can, or should, decide the degree of relative importance. Which is more significant, a breakthrough in basic research, which years later may lead to an effective cure for certain cancers, or a new formulation that enables patients to administer an old but effective medication themselves instead of having to visit a physician or a hospital three times a week? It is impossible to decide. Equally, a company must choose between a new way to service customers, which enables the company to retain an important account it would otherwise have lost, and a new product, which gives the company leadership in markets that, while still small, may within a few years become big and important ones. These are judgments rather than measurements. But they are not arbitrary; they are not even subjective. And they are quite rigorous even though not capable of quantification. Above all, they do what a "measurement" is meant to enable us to do: to take purposeful action based on knowledge rather than on opinion or guesswork.

The most important question for the typical business in this review is probably: Have we gained innovative leadership, or at least maintained it? Leadership does not necessarily equate with size. It means to be accepted as the leader, recognized as the standard-setter; above all, it means having the freedom to lead rather than being obliged to follow. This is the acid test of successful entrepreneurship in the existing business.

V

STRUCTURES

Policies, practices, and measurements make possible entrepreneurship and innovation. They remove or reduce possible impediments. They create the proper attitude and provide the proper tools. But innovation is done by people. And people work within a structure.

For the existing business to be capable of innovation, it has to create a structure that allows people to be entrepreneurial. It has to devise relationships that center on entrepreneurship. It has to make sure that its rewards and incentives, its compensation, personnel decisions, and policies, all reward the right entrepreneurial behavior and do not penalize it.

1. This means, first, that the entrepreneurial, the new, has to be organized separately from the old and existing. Whenever we have tried to make an existing unit the carrier of the entrepreneurial project,

we have failed. This is particularly true, of course, in the large business, but it is true in medium-sized businesses as well, and even in small businesses.

One reason is that (as said earlier) the existing business always requires time and effort on the part of the people responsible for it, and deserves the priority they give it. The new always looks so puny—so unpromising—next to the reality of the massive, ongoing business. The existing business, after all, has to nourish the struggling innovation. But the "crisis" in today's business has to be attended to as well. The people responsible for an existing business will therefore always be tempted to postpone action on anything new, entrepreneurial, or innovative until it is too late. No matter what has been tried—and we have now been trying every conceivable mechanism for thirty or forty years—existing units have been found to be capable mainly of extending, modifying, and adapting what already is in existence. The new belongs elsewhere.

2. This means also that there has to be a special locus for the new venture within the organization, and it has to be pretty high up. Even though the new project, by virtue of its current size, revenues, and markets, does not rank with existing products, somebody in top management must have the specific assignment to work on tomorrow as an entrepreneur and innovator.

This need not be a full-time job; in the smaller business, it very often cannot be a full-time job. But it needs to be a clearly defined job and one for which somebody with authority and prestige is fully accountable. These people will normally also be responsible for the policies necessary to build entrepreneurship into the existing business, for the abandonment analysis, for the Business X-Ray, and for developing the innovation objectives to plug the gap between what can be expected of the existing products and services and what is needed for survival and growth of the company. They are also normally charged with the systematic analysis of innovative opportunities—the analysis of the innovative opportunities presented in the preceding section of this book, the Practice of Innovation. They should be further charged with responsibility for the analysis of the innovative and entrepreneurial ideas that come up from the organization, for example, in the recommended "informal" session with the juniors.

And innovative efforts, especially those aimed at developing new businesses, products, or services, should normally report directly to this "executive in charge of innovation" rather than to managers further down the hierarchy. They should never report to line managers charged with responsibility for ongoing operations.

This will be considered heresy in most companies, particularly "well-managed" ones. But the new project is an infant and will remain one for the foreseeable future, and infants belong in the nursery. The "adults," that is, the executives in charge of existing businesses or products, will have neither time nor understanding for the infant project. They cannot afford to be bothered.

Disregard of this rule cost a major machine-tool manufacturer its leadership in robotics.

The company had the basic patents on machine tools for automated mass production. It had excellent engineering, an excellent reputation, and first-rate manufacturing. Everyone in the early years of factory automation—around 1975—expected it to emerge as the leader. Ten years later it had dropped out of the race entirely. The company had placed the unit charged with the development of machine tools for automated production three or four levels down in the organization, and had it report to people charged with designing, making, and selling the company's traditional machine-tool lines. These people were supportive; in fact, the work on robotics had been mainly their idea. But they were far too busy defending their traditional lines against a lot of new competitors such as the Japanese, redesigning them to fit new specifications, demonstrating, marketing, financing, and servicing them. Whenever the people in charge of the "infant" went to their bosses for a decision, they were told, "I have no time now, come back next week." Robotics were, after all, only a promise; the existing machine-tool lines produced millions of dollars each year.

Unfortunately, this is a common error.

The best, and perhaps the only, way to avoid killing off the new by sheer neglect is to set up the innovative project from the start as a separate business.

The best known practitioners of this approach are three American companies: Procter & Gamble, the soap, detergent, edible oil, and food producer—a very large and aggressively entrepreneurial company; Johnson & Johnson, the hygiene and health-care supplier; and 3M, a major manufacturer of industrial and consumer products. These three companies differ in the details of practice but essentially all three have the same policy. They set up the new venture as a separate business from the beginning and put a project manager in charge. The project manager remains in charge until the project is either abandoned or has achieved its objective and become a full-fledged business. And until then, the project manager can mobilize all the skills as they are needed—research, manufacturing, finance, marketing—and put them to work on the project team.

A company that engages in more than one innovative effort at a time (and bigger companies usually do) might have all the "infants" report directly to the same member of the top management group. It does not greatly matter that the ventures have different technologies, markets, or product characteristics. They all are new, small, and entrepreneurial. They are all exposed to the same "childhood diseases." The problems from which the entrepreneurial venture suffers, and the decisions it requires, tend to be pretty much the same regardless of technology, of market, or of product line. Somebody has to have time for them, to give them the attention they need, to take the trouble to understand what the problems are, the crucial decisions, the things that really matter in a given innovative effort. And this person has to have sufficient stature in the business to be able to represent the infant project—and to make the decision to stop an effort if it is going nowhere.

3. There is another reason why a new, innovative effort is best set up separately: to keep away from it the burdens it cannot yet carry. Both the investment in a new product line and its returns should, for instance, not be included in the traditional return-on-investment analysis until the product line has been on the market for a number of years. To ask the fledgling development to shoulder the full burdens an existing business imposes on its units is like asking a six-year-old to go on a long hike carrying a sixty-pound pack; neither will get very far. And yet the existing business has requirements with respect to accounting, to personnel policy, to reporting of all kinds, which it cannot easily waive.

The innovative effort and the unit that carries it require different policies, rules, and measurements in many areas. How about the company's pension plan, for instance? Often it makes sense to give people in the innovative unit a participation in future profits rather than to put them into a pension plan when they are producing, as yet, no earnings to supply a pension fund contribution.

The area in which separation of the new, innovative unit from the ongoing business is most important is compensation and rewards of key people. What works best in a going, established business would kill the "infant"—and yet not be adequate compensation for its key people.

Indeed, the compensation scheme that is most popular in large businesses, one based on return on assets or on investment, is a near-complete bar to innovation.

I learned this many years ago in a major chemical company. Everybody knew that one of its central divisions had to produce new materials to stay in business. The plans for these materials were there, the scientific work had been done ... but nothing happened. Year after year there was another excuse. Finally, the division's general manager spoke up at a review meeting, "My management group and I are compensated primarily on the basis of return-on-investment. The moment we spend money on developing the new materials, our return will go down by half for at least four years. Even if I am still here in four years time when we should show the first returns on these investments—and I doubt that the company will put up with me that long if profits are that much lower—I'm taking bread out of the mouths of all my associates in the meantime. Is it reasonable to expect us to do this?" The formula was changed and the developmental expenses for the new project were taken out of the return-on-investment figures. Within eighteen months the new materials were on the market. Two years later they had given the division leadership in its field which it has retained to this day. Four years later the division doubled its profits.

In terms of compensation and rewards for innovative efforts, however, it is far easier to define what should not be done than it is to spell out what should. The requirements are conflicting: the new project must not be burdened with a compensation load it cannot carry, yet the people involved must be adequately motivated by rewards appropriate to their efforts.

Specifically, this means that the people in charge of the new project should be kept at a moderate salary. It is, however, quite unrealistic to ask them to work for less money than they received in their old jobs. People put in charge of a new area within an existing business are likely to make good money. They are also the people who could easily move to other jobs, either within or outside the company, in which they would make more money. One therefore has to start out with their existing compensation and benefits.

One method that both 3M and Johnson & Johnson use effectively is to promise that the person who successfully develops a new product, a new market, or a new service and then builds a business on it will become the head of that business: general manager, vice-president, or division president, with the rank, compensation, bonuses, and stock options appropriate to the level. This can be a sizable reward, and yet it does not commit the company to anything except in case of success.

Another method—and which one is preferable will depend largely on the tax laws at the time—is to give the people who take on the new development a share in future profits. The venture might, for instance, be treated as if it were a separate company in which the entrepreneurial managers in charge have a stake, say 25 percent. When the venture reaches maturity, they are bought out at a pre-set formula based on sales and profits.

One thing more is needed: the people who take on the innovating task in an existing business also "venture." It is only fair that their employer share the risk. They should have the option of returning to their old job at their old compensation rate if the innovation fails. They should not be rewarded for failure, but they should certainly not be penalized for trying.

4. As implied in discussing individual compensation, the returns on innovation will be quite different from those of the existing business and will have to be measured differently. To say, "We expect all our businesses to show at least a fifteen percent pre-tax return each year and ten percent annual growth" may make sense for existing businesses and existing products. It makes absolutely no sense for the new project, being at once much too high and much too low.

For a long time (years, in many cases) the new endeavor shows neither profits nor growth. It absorbs resources. But then it should grow very fast for quite a long time and return the money invested in its development at least fifty-fold—if not at a much higher rate—or else the innovation is a failure. An innovation starts small but it should end big. It should result in a new major business rather than in just another "specialty" or a "respectable" addition to the product line.

Only by analyzing a company's own innovative experience, the feedback from its performance on its expectations, can the company determine what the appropriate expectations are for innovations in its industry and its markets. What are the appropriate time spans? And what is the optimal distribution of effort? Should there be a heavy investment of men and money at the beginning, or should the effort at the start be confined to one person, with a helper or two, working alone? When should the effort then be scaled up? And when should "development" become "business," producing large but conventional returns?

These are key questions. The answers to them are not to be found in books. Yet they cannot be answered arbitrarily, by hunch, or by fighting it out. Entrepreneurial companies do know what patterns, rhythms, and time spans pertain to innovations in their specific industry, technology, and market.

The innovative major bank mentioned earlier knows, for instance, that a new subsidiary established in a new country will require investment for at least three years. It should break even in the fourth year, and should have repaid the total investment by the middle of the sixth year. If it still requires investment by the end of the sixth year, it is a disappointment and should probably be shut down.

A new major service—leasing, for example—has a similar though somewhat shorter cycle. Procter & Gamble—or so it looks from the outside—knows that its new products should be on the market and selling two to three years after work on them has begun. They should have established themselves as market leaders eighteen months later. IBM, it seems, figures on a five-year lead time for a new major product before market introduction. Within another year the new product should then start to grow fast. It should attain market leadership and profitability fairly early in its second year on the market, have repaid the full investment by the early months of the third year, and peak and level out in its fifth year on the market. By then, a new IBM product should already have begun to make it obsolescent.

The only way, however, to know these things is through the systematic analysis of the performance of the company and of its competitors, that is, by systematic feedback from innovation results to innovation expectations and by regular appraisal of the company's performance as entrepreneur.

And once a company understands what results should and could be expected from its innovative efforts, it can then design the appropriate controls. These will both measure how well units and their managers perform in innovation and determine which innovative efforts to push, which to reconsider, and which to abandon.

5. The final structural requirement for entrepreneurship in the existing business is that a person or a component group should be held clearly accountable.

In the "middle-sized growth companies" mentioned earlier, this is usually the primary responsibility of the chief executive officer (CEO). In large companies, it probably is more likely a designated and very senior member of the top management group. In smaller businesses, this executive in charge of entrepreneurship and innovation may well carry other responsibilities as well.

The cleanest organizational structure for entrepreneurship, though suitable only in the very large company, is a totally separate innovating operation or development company.

The earliest example of this was set up more than one hundred years ago, in 1872, by Hefner-Alteneck, the first college-trained engineer hired by a manufacturing company anywhere, the German Siemens Company. Hefner started the first "research lab" in industry. Its members were charged with inventing new and different products and processes. But they were also responsible for identifying new and different end uses and new and different markets. And they not only did the technical work; they were responsible for development of the manufacturing process, for the introduction of the new product into the marketplace, and for its profitability.

Fifty years later, in the 1920s, the American DuPont Company independently set up a similar unit and called it a Development Department. This department gathers innovative ideas from all over the company, studies them, thinks them through, analyzes them. Then it proposes to top management which ones should be tackled as major innovative projects. From the beginning, it brings to bear on the innovation all the resources needed: research, development, manufacturing, marketing, finance, and so on. It is in charge until the new product or service has been on the market for a few years.

Whether the responsibility for innovation rests with the chief executive officer, with another member of top management, or with a separate component, whether it is a full-time assignment or part of an executive's responsibilities, it should always be set up and recognized both as a separate responsibility and as a responsibility of top management. And it should always include the systematic and purposeful search for innovative opportunities.

It might be asked, Are all these policies and practices necessary? Don't they interfere with the entrepreneurial spirit and stifle creativity? And cannot a business be entrepreneurial without such policies and practices? The answer is, Perhaps, but neither very successfully nor for very long.

Discussions of entrepreneurship tend to focus on the personalities and attitudes of top management people, and especially of the chief executive.* Of course, any top management can damage and stifle entrepreneurship within its company. It's easy enough. All it takes is to say "No" to every new idea and to keep on saying it for a few years—and then make sure that those who came up with the new ideas never get a reward or a promotion and become ex-employees fairly swiftly. It is far less certain, however, that top management personalities and attitudes can by themselves—without the proper policies and practices—create an entrepreneurial business, which is what most of the books on entrepreneurship assert, at least by implication. In the few short-lived cases I know of, the companies were built and still run by the founder. Even then, when it gets to be successful the company soon ceases to be entrepreneurial unless it adopts the policies and practices of entrepreneurial management. The reason why top management personalities and attitudes do not suffice in any but the very young or very small business is, of course, that even a medium-sized enterprise is a pretty large organization. It requires a good many people who know what they are supposed to do, want to do it, are motivated toward doing it, and are supplied with both the tools and continuous reaffirmation. Otherwise there is only lip service; entrepreneurship soon becomes confined to the CEO's speeches.

And I know of no business that continued to remain entrepreneurial beyond the founder's departure, unless the founder had built into the organization the policies and practices of entrepreneurial management. If these are lacking, the business becomes timid and backward-looking within a few years at the very latest. And these companies do not even realize, as a rule, that they have lost their essential quality, the one element that had made them stand out, until it is perhaps too late. For this realization one needs a measurement of entrepreneurial performance.

Two companies that were entrepreneurial businesses *par excellence* under their founders' management are good examples: Walt Disney Productions and McDonald's. The respective founders, Walt Disney and Ray Kroc, were men of tremendous imagination and drive, each the very embodiment of creative, entrepreneurial, and innovative thinking. Both built into their companies strong operating day-to-day

*The best presentation of this viewpoint is in Rosabeth M. Kanter's *The Change Masters* (New York: Simon & Schuster, 1983).

management. But both kept to themselves the entrepreneurial responsibility within their companies. Both depended on the "entrepreneurial personality" and did not embed the entrepreneurial spirit in specific policies and practices. Within a few years after the death of these men, their companies had become stodgy, backward-looking, timid, and defensive.

Companies that have built entrepreneurial management into their structure—Procter & Gamble, Johnson & Johnson, Marks and Spencer —continue to be innovators and entrepreneurial leaders decade after decade, irrespective of changes in chief executives or economic conditions.

VI

STAFFING

How should the existing business staff for entrepreneurship and innovation? Are there such people as "entrepreneurs"? Are they a special breed?

The literature is full of discussions of these questions; full of stories of the "entrepreneurial personality" and of people who will never do anything but innovate. In the light of our experience—and it is considerable—these discussions are pointless. By and large, people who do not feel comfortable as innovators or as entrepreneurs will not volunteer for such jobs; the gross misfits eliminate themselves. The others can learn the practice of innovation. Our experience shows that an executive who has performed in other assignments will do a decent job as an entrepreneur. In successful entrepreneurial businesses, nobody seems to worry whether a given person is likely to do a good job of development or not. People of all kinds of temperaments and backgrounds apparently do equally well. Any young engineer in 3M who comes to top management with an idea that makes sense is expected to take on its development.

Equally, there is no reason to worry where the successful entrepreneur will end up. To be sure, there are some people who only want to work on new projects and never want to run anything. When most English families still had nannies, many did not want to stay after "their" baby got to the stage when it began to walk and talk—in other words, when it was no longer a baby. But many were perfectly content to stay on and did not find it difficult to look after a much older child. The people who do not want to be anything but entrepreneurs are unlikely to be in the employ of an existing business to begin with, and even more unlikely to have been successful in it. And the people who do well as entrepreneurs in an existing business have, as a rule, proved themselves earlier as managers in the same organization. It is thus reasonable to assume that they can both innovate and manage what already exists. There are some people at Procter & Gamble and at 3M who make a career of being project managers and who take on a new project as soon as they have successfully finished an old one. But most people at the higher levels of these companies have made their careers out of "project management," into "product management," into "market management," and finally into a senior company-wide position. And the same is true of Johnson & Johnson and of Citibank.

The best proof that entrepreneurship is a question of behavior, policies, and practices rather than personality is the growing number of older large-company people in the United States who make entrepreneurship their second career. Increasingly, middle- and upper-level executives and senior professionals who have spent their entire working lives in large companies—more often than not with the same employer—take early retirement after twenty-five or thirty years of service when they have reached what they realize is their terminal job. At fifty or fifty-five, these middle-aged people then become entrepreneurs. Some start their own business. Some, especially technical specialists, set up shop as consultants to new and small ventures. Some join a new small company in a senior position. And the great majority are both successful and happy in their new assignment.

Modern Maturity, the magazine of the American Association of Retired Persons, is full of stories of such people, and of advertisements by new small companies looking for them. In a management seminar for chief executive officers that I ran in 1983, there were fifteen such second-career entrepreneurs (fourteen men and one woman) among the forty-eight participants. During a special session for these people, I asked them whether they had been frustrated or stifled while working all those years for big companies, as "entrepreneurial personalities" are supposed to be. They thought the question totally absurd. I then asked whether they had much difficulty changing their roles; they thought this equally absurd. As one of them said—and all the others nodded assent—"Good management is good management, whether you run a

$180 million department at General Electric, with its billions of sales as I used to do, or a new, growing diagnostic-instrument innovator with $6 million in sales, as I do now. Of course I do different things and do things differently. But I apply the concepts I learned at G.E. and do exactly the same analysis. The transition was easier, in fact, than when I moved, ten years earlier, from being a bench engineer into my first management job."

Public-service institutions teach the same lesson. Among the most successful innovators in recent American history are two men in higher education, Alexander Schure and Ernest Boyer. Schure started out as a successful inventor in the electronics field, with a good many patents to his name. But in 1955, when he was in his early thirties, he founded the New York Institute of Technology as a private university without support from government, foundation, or big company, and with brand-new ideas regarding the kind of students to be recruited and what they were to be taught as well as how. Thirty years later, his institute has become a leading technical university with four campuses, one of them a medical school, and almost twelve thousand students. Schure still works as a successful electronics inventor. But he has also been for these thirty years the full-time chancellor of his university, and has, by all accounts, built up a professional and effective management team.

In contrast to Schure, Boyer started out as an administrator, first in the University of California system, then in the State University of New York, which with 350,000 student and 64 campuses is the biggest and most bureaucratic of American university systems. By 1970, Boyer, at forty-two, had worked his way to the top and was appointed chancellor. He immediately founded the Empire State College—actually not a college at all but an unconventional solution to one of the oldest and most frustrating failures of American higher education, the degree program for adults who do not have full academic credentials.

Although tried many times, this had never worked before. If these adults were admitted to college programs together with the "regular" younger students, no attention was usually paid to their aims, their needs, and least of all to their experience. They were treated as if they were eighteen years old, got discouraged, and soon dropped out. But if, as was tried repeatedly, they were put into special "continuing education programs," they were likely to be considered a nuisance and shoved aside, with programs staffed by whatever faculty the university could most easily spare. In Boyer's Empire State College, the adults attend regular university courses in one of the colleges or universities of the state university. But first the adult students are assigned a "mentor," usually a member of a nearby state university faculty. The mentor helps them work out their programs and decide whether they need special preparation, and where, conversely, their experience qualifies them for advanced standing and work. And then the mentor acts as broker, negotiating admission, standing, and program for each applicant with the appropriate institution.

All this may sound like common sense—and so it is. Yet it was quite a break with the habits and mores of American academia and was fought hard by the state university establishment. But Boyer persisted. His Empire State College program has now become the first successful program of this kind in American higher education, with more than six thousand students, a negligible dropout rate, and a master's program. Boyer, the arch-innovator, did not cease to be an "administrator." From chancellor of the State University of New York he went on to become, first, President Carter's Commissioner of Education, and then president of the Carnegie Foundation for the Advancement of Teaching—respectively, the most "bureaucratic" and the most "establishment" job in American academia.

These examples do not prove that anyone can excel at being both a bureaucrat and an innovator. Schure and Boyer are surely exceptional people. But their experiences do show that there is no specific "personality" for either task. What is needed is willingness to learn, willingness to work hard and persistently, willingness to exercise self-discipline, willingness to adapt and to apply the right policies and practices. Which is exactly what any enterprise that adopted entrepreneurial management has found out with respect to people and staffing.

To enable the entrepreneurial project to be run successfully, as something new, the structure and organization have to be right; relationships have to be appropriate; and compensation and rewards have to fit. But when all this has been done, the question of who is to run the unit, and what should be done with them when they have succeeded in building up the new project, must be decided on an individual basis for this person or that person, rather than according to this or that psychological theory for none of which there is much empirical evidence.

Staffing decisions in the entrepreneurial business are made like any other decision about people and jobs. Of course, they are risk-taking decisions: decisions about people always are. Of course, they have to be made carefully and conscientiously. And they have to be made the correct way. First, the assignment must be thought through; then one considers a number of people; then one checks carefully their performance records; and finally one checks out each of the candidates with a few people for whom he or she has worked. But all this applies to every decision that puts a person into a job. And in the entrepreneurial company, the batting average in people-decisions is the same for entrepreneurs as it is for other managerial and professional people.

VII

THE DONT'S

There are some things the entrepreneurial management of an existing business should not do.

1. The most important caveat is not to mix managerial units and entrepreneurial ones. Do not ever put the entrepreneurial into the existing managerial component. Do not make innovation an objective for people charged with running, exploiting, optimizing what already exists.

But it is also inadvisable—in fact, almost a guarantee of failure—for a business to try to become entrepreneurial without changing its basic policies and practices. To be an entrepreneur on the side rarely works.

In the last ten or fifteen years a great many large American companies have tried to go into joint ventures with entrepreneurs. Not one of these attempts has succeeded; the entrepreneurs found themselves stymied by policies, by basic rules, by a "climate" they felt was bureaucratic, stodgy, reactionary. But at the same time their partners, the people from the big company, could not figure out what the entrepreneurs were trying to do and thought them undisciplined, wild, visionary.

By and large, big companies have been successful as entrepreneurs only if they use their own people to build the venture. They have been successful only when they use people whom they understand and who understand them, people whom they trust and who in turn know how to get things done in the existing business; people, in other words, with whom one can work as partners. But this presupposes that the entire company is imbued with the entrepreneurial spirit, that it wants innovation and is reaching out for it, considering it both a necessity and an opportunity. It presupposes that the entire organization has been made "greedy for new things."

2. Innovative efforts that take the existing business out of its own field are rarely successful. Innovation had better not be "diversification." Whatever the benefits of diversification, it does not mix with entrepreneurship and innovation. The new is always sufficiently difficult not to attempt it in an area one does not understand. An existing business innovates where it has expertise, whether knowledge of market or knowledge of technology. Anything new will predictably get into trouble, and then one has to know the business. Diversification itself rarely works unless it, too, is built on commonality with the existing business, whether commonality of the market or commonality of the technology. Even then, as I have discussed elsewhere,* diversification has its problems. But if one adds to the difficulties and demands of diversification the difficulties and demands of entrepreneurship, the result is predictable disaster. So one innovates only where one understands.

3. Finally, it is almost always futile to avoid making one's own business entrepreneurial by "buying in," that is, by acquiring small entrepreneurial ventures. Acquisitions rarely work unless the company that does the acquiring is willing and able within a fairly short time to furnish management to the acquisition. The managers that have come with the acquired company rarely stay around very long. If they were owners, they have now become wealthy; if they were professional managers, they are likely to stay around only if given much bigger opportunities in the new, acquiring company. So, within a year or two, the acquirer has to furnish management to run the business that has been bought. This is particularly true when a non-entrepreneurial company buys an entrepreneurial one. The management people in the new acquired venture soon find that they cannot work with the people in their new parent company, and vice versa. I myself know of no case where "buying in" has worked.

A business that wants to be able to innovate, wants to have a chance to succeed and prosper in a time of rapid change, has to build entre-

*In *Management: Tasks, Responsibilities, Practices*, especially Chapters 56 and 57.

preneurial management into its own system. It has to adopt policies that create throughout the entire organization the desire to innovate and the habits of entrepreneurship and innovation. To be a successful entrepreneur, the existing business, large or small, has to be managed as an entrepreneurial business.

14

Entrepreneurship in the Service Institution

Public-service institutions such as government agencies, labor unions, churches, universities, and schools, hospitals, community and charitable organizations, professional and trade associations and the like, need to be entrepreneurial and innovative fully as much as any business does. Indeed, they may need it more. The rapid changes in today's society, technology, and economy are simultaneously an even greater threat to them and an even greater opportunity.

Yet public-service institutions find it far more difficult to innovate than even the most "bureaucratic" company. The "existing" seems to be even more of an obstacle. To be sure, every service institution likes to get bigger. In the absence of a profit test, size is the one criterion of success for a service institution, and growth a goal in itself. And then, of course, there is always so much more that needs to be done. But stopping what has "always been done" and doing something new are equally anathema to service institutions, or at least excruciatingly painful to them.

Most innovations in public-service institutions are imposed on them either by outsiders or by catastrophe. The modern university, for instance, was created by a total outsider, the Prussian diplomat Wilhelm von Humboldt. He founded the University of Berlin in 1809 when the traditional university of the seventeenth and eighteenth century had been all but completely destroyed by the French Revolution and the Napoleonic wars. Sixty years later, the modern American university came into being when the country's traditional colleges and universities were dying and could no longer attract students.

Similarly, all basic innovations in the military in this century, whether in structure or in strategy, have followed on ignominious malfunction or crushing defeat: the organization of the American Army and of its strategy by a New York lawyer, Elihu Root, Teddy Roosevelt's Secretary of War, after its disgraceful performance in the Spanish-American War; the reorganization, a few years later, of the British Army and its strategy by Secretary of War Lord Haldane, another civilian, after the equally disgraceful performance of the British in the Boer War; and the rethinking of the German Army's structure and strategy after the defeat of World War I.

And in government, the greatest innovative thinking in recent political history, America's New Deal of 1933–36, was triggered by a Depression so severe as almost to unravel the country's social fabric.

Critics of bureaucracy blame the resistance of public-service institutions to entrepreneurship and innovation on "timid bureaucrats," on time-servers who "have never met a payroll," or on "power-hungry politicians." It is a very old litany—in fact, it was already hoary when Machiavelli chanted it almost five hundred years ago. The only thing that changes is who intones it. At the beginning of this century, it was the slogan of the so-called liberals and now it is the slogan of the so-called neo-conservatives. Alas, things are not that simple, and "better people"—that perennial panacea of reformists—are a mirage. The most entrepreneurial, innovative people behave like the worst time-serving bureaucrat or power-hungry politician six months after they have taken over the management of a public-service institution, particularly if it is a government agency.

The forces that impede entrepreneurship and innovation in a public-service institution are inherent in it, integral to it, inseparable from it.* The best proof of this are the internal staff services in businesses, which are, in effect, the "public-service institutions" within business corporations. These are typically headed by people who have come out of operations and have proven their capacity to perform in competitive markets. And yet the internal staff services are not notorious as innovators. They are good at building empires—and they always want to do more of the same. They resist abandoning anything they are doing. But they rarely innovate once they have been established.

*On the public-service institution and its characteristics, see the section on Performance in the Service Institution, Chapters 11–14, in *Management: Tasks, Responsibilities, Practices.*

There are three main reasons why the existing enterprise presents so much more of an obstacle to innovation in the public-service institution than it does in the typical business enterprise.

1. First, the public-service institution is based on a "budget" rather than being paid out of its results. It is paid for its efforts and out of funds somebody else has earned, whether the taxpayer, the donors of a charitable organization, or the company for which a personnel department or the marketing services staff work. The more efforts the public service institution engages in, the greater its budget will be. And "success" in the public-service institution is defined by getting a larger budget rather than obtaining results. Any attempt to slough off activities and efforts therefore diminishes the public-service institution. It causes it to lose stature and prestige. Failure cannot be acknowledged. Worse still, the fact that an objective has been attained cannot be admitted.

2. Second, a service institution is dependent on a multitude of constituents. In a business that sells its products on the market, one constituent, the consumer, eventually overrides all the others. A business needs only a very small share of a small market to be successful. Then it can satisfy the other constituents, whether shareholders, workers, the community, and so on. But precisely because public-service institutions —and that includes the staff activities within a business corporation— have no "results" out of which they are being paid, any constituent, no matter how marginal, has in effect a veto power. A public-service institution has to satisfy everyone; certainly, it cannot afford to alienate anyone.

The moment a service institution starts an activity, it acquires a "constituency," which then refuses to have the program abolished or even significantly modified. But anything new is always controversial. This means that it is opposed by existing constituencies without having formed, as yet, a constituency of its own to support it.

3. The most important reason, however, is that public-service institutions exist after all to "do good." This means that they tend to see their mission as a moral absolute rather than as economic and subject to a cost/benefit calculus. Economics always seeks a different allocation of the same resources to obtain a higher yield. Everything economic is therefore relative. In the public-service institution, there is no such thing as a higher yield. If one is "doing good," then there is no "better."

Indeed, failure to attain objectives in the quest for a "good" only means that efforts need to be redoubled. The forces of evil must be far more powerful than expected and need to be fought even harder.

For thousands of years the preachers of all sorts of religions have held forth against the "sins of the flesh." Their success has been limited, to say the least. But this is no argument as far as the preachers are concerned. It does not persuade them to devote their considerable talents to pursuits in which results may be more easily attainable. On the contrary, it only proves that their efforts need to be redoubled. Avoiding the "sins of the flesh" is clearly a "moral good," and thus an absolute, which does not admit of any cost/benefit calculation.

Few public-service institutions define their objectives in such absolute terms. But even company personnel departments and manufacturing service staffs tend to see their mission as "doing good," and therefore as being moral and absolute instead of being economic and relative.

This means that public-service institutions are out to maximize rather than to optimize. "Our mission will not be completed," asserts the head of the Crusade Against Hunger, "as long as there is one child on the earth going to bed hungry." If he were to say, "Our mission will be completed if the largest possible number of children that can be reached through existing distribution channels get enough to eat not to be stunted," he would be booted out of office. But if the goal is maximization, it can never be attained. Indeed, the closer one comes toward attaining one's objective, the more efforts are called for. For, once optimization has been reached (and the optimum in most efforts lies between 75 and 80 percent of theoretical maximum), additional costs go up exponentially while additional results fall off exponentially. The closer a public-service institution comes to attaining its objectives, therefore, the more frustrated it will be and the harder it will work on what it is already doing.

It will, however, behave exactly the same way the less it achieves. Whether it succeeds or fails, the demand to innovate and to do something else will be resented as an attack on its basic commitment, on the very reason for its existence, and on its beliefs and values.

These are serious obstacles to innovation. They explain why, by and large, innovation in public services tends to come from new ventures rather than from existing institutions.

The most extreme example around these days may well be the labor union. It is probably the most successful institution of the century in the developed countries. It has clearly attained its original objectives. There can be no more "more" when the labor share of gross national product in Western developed countries is around 90 percent—and in some countries, such as Holland, close to 100 percent. Yet the labor union is incapable of even thinking about new challenges, new objectives, new contributions. All it can do is repeat the old slogans and fight the old battles. For the "cause of labor" is an absolute good. Clearly, it must not be questioned, let alone redefined.

The university, however, may not be too different from the labor union, and in part for the same reason—a level of growth and success second in this century only to that of the labor union.

Still there are enough exceptions among public-service institutions (although, I have to admit, not many among government agencies) to show that public-service institutions, even old and big ones, can innovate.

One Roman Catholic archdiocese in the United States, for instance, has brought in lay people to run the diocese, including a married lay woman, the former personnel vice-president of a department store chain, as the general manager. Everything that does not involve dispensing sacraments and ministering to congregations is done by lay professionals and managers. Although there is a shortage of priests throughout the American Catholic Church, this archdiocese has priests to spare and has been able to move forward aggressively to build congregations and expand religious services.

One of the oldest of scientific societies, the American Association for the Advancement of Science, redirected itself between 1960 and 1980 to become a "mass organization" without losing its character as a leader. It totally changed its weekly magazine, *Science,* to become the spokesman for science to public and government, and to be the authoritative reporter on science policy. And it created a scientifically solid yet popular mass circulation magazine for lay readers.

A large hospital on the West Coast recognized, as early as 1965 or so, that health care was changing as a result of its success. Where other large city hospitals tried to fight such trends as those toward hospital chains or freestanding ambulatory treatment centers, this institution has been an innovator and a leader in these developments. Indeed, it was the first to build a freestanding maternity center in which the expectant mother is given a motel room at fairly low cost, yet with all the medical services available should they be needed. It was the first to go into freestanding surgical centers for ambulatory care. But it also started to build its own voluntary hospital chain, in which it offers management contracts to smaller hospitals throughout the region.

Beginning around 1975, the Girl Scouts of the U.S.A., a large organization dating back to the early years of the century with several million young women enrolled, introduced innovations affecting membership, programs, and volunteers—the three basic dimensions of the organization. It began actively to recruit girls from the new urban middle classes, that is, blacks, Asians, Latins; these minorities now account for one-fifth of the members. It recognized that with the movement of women into professions and managerial positions, girls need new programs and role models that stress professional and business careers rather than the traditional careers as homemaker or nurse. The Girl Scouts management people realized that the traditional sources for volunteers to run local activities were drying up because young mothers no longer were sitting at home searching for things to do. But they recognized, too, that the new professional, the new working mother represents an opportunity and that the Girl Scouts have something to offer her; and for any community organization, volunteers are the critical constraint. They therefore set out to make work as a volunteer for the Girl Scouts attractive to the working mother as a good way to have time and fun with her child while also contributing to her child's development. Finally, the Girl Scouts realized that the working mother who does not have enough time for her child represents another opportunity: they started Girl Scouting for preschool children. Thus, the Girl Scouts reversed the downward trend in enrollment of both children and volunteers, while the Boy Scouts—a bigger, older, and infinitely richer organization—is still adrift.

II

ENTREPRENEURIAL POLICIES

These are all American examples, I fully realize. Doubtless, similar examples are to be found in Europe or Japan. But I hope that these cases, despite their limitations, will suffice to demonstrate the entrepreneurial policies needed in the public-service institution to make it capable of innovation.

1. First, the public-service institution needs a clear definition of its mission. What is it trying to do? Why does it exist? It needs to focus on objectives rather than on programs and projects. Programs and projects are means to an end. They should always be considered as temporary and, in fact, short-lived.

2. The public-service institution needs a realistic statement of goals. It should say, "Our job is to assuage famine," rather than, "Our job is to eliminate hunger." It needs something that is genuinely attainable and therefore a commitment to a realistic goal, so that it can say eventually, "Our job is finished."

There are, of course, objectives that can never be attained. To administer justice in any human society is clearly an unending task, one that can never be fully accomplished even to modest standards. But most objectives can and should be phrased in optimal rather than in maximal terms. Then it is possible to say: "We have attained what we were trying to do."

Surely, this should be said with respect to the traditional goals of the schoolmaster: to get everyone to sit in school for long years. This goal has long been attained in developed countries. What does education have to do now, that is, what is the meaning of "education" as against mere schooling?

3. Failure to achieve objectives should be considered an indication that the objective is wrong, or at least defined wrongly. The assumption has then to be that the objective should be economic rather than moral. If an objective has not been attained after repeated tries, one has to assume that it is the wrong one. It is not rational to consider failure a good reason for trying again and again. The probability of success, as mathematicians have known for three hundred years, diminishes with each successive try; in fact, the probability of success in any succeeding try is never more than one-half the probability of the preceding one. Thus, failure to attain objectives is a *prima facie* reason to question the validity of the objective—the exact opposite of what most public-service institutions believe.

4. Finally, public-service institutions need to build into their policies and practices the constant search for innovative opportunity. They need to view change as an opportunity rather than a threat.

The innovating public-service institutions mentioned in the preceding pages succeeded because they applied these basic rules.

In the years after World War II, the Roman Catholic Church in the

United States was confronted for the first time with the rapid emergence of a well-educated Catholic laity. Most Catholic dioceses, and indeed most institutions of the Roman Catholic Church, perceived in this a threat, or at least a problem. With an educated Catholic laity, unquestioned acceptance of bishop and priest could no longer be taken for granted. And yet there was no place for Catholic lay people in the structure and governance of the Church. Similarly, all Roman Catholic dioceses in the United States, beginning around 1965 or 1970, faced a sharp drop in the number of young men entering the priesthood—and perceived this as a major threat. Only one Catholic archdiocese saw both as opportunities. (As a result, it has a different problem. Young priests from all over the United States want to enter it; for in this one archdiocese, the priest gets to do the things he trained for, the things which he entered the priesthood to do.)

All American hospitals, beginning in 1970 or 1975, saw changes coming in the delivery of health care. Most of them organized themselves to fight these changes. Most of them told everybody that "these changes will be catastrophic." Only the one hospital saw in them opportunities.

The American Association for the Advancement of Science saw in the expansion of people with scientific backgrounds and working in scientific pursuits a tremendous opportunity to establish itself as a leader, both within the scientific community and outside.

And the Girl Scouts looked at demographics and said: "How can we convert population trends into new opportunities for us?"

Even in government, innovation is possible if simple rules are obeyed. Here is one example.

Lincoln, Nebraska, 120 years ago, was the first city in the Western world to take into municipal ownership public services such as public transportation, electric power, gas, water, and so on. In the last ten years, under a woman mayor, Helen Boosalis, it has begun to privatize such services as garbage pickup, school transportation, and a host of others. The city provides the money, with private businesses bidding for the contracts; there are substantial savings in cost and even greater improvements in service.

What Helen Boosalis has seen in Lincoln is the opportunity to separate the "provider" of public services, that is, government, and the "supplier." This makes possible both high service standards and the efficiency, reliability, and low cost which competition can provide.

The four rules outlined above constitute the *specific* policies and practices the public-service institution requires if it is to make itself entrepreneurial and capable of innovation. In addition, however, it also needs to adopt those policies and practices that any existing organization requires in order to be entrepreneurial, the policies and practices discussed in the preceding chapter, The Entrepreneurial Business.

III

THE NEED TO INNOVATE

Why is innovation in the public-service institution so important? Why cannot we leave existing public-service institutions the way they are, and depend for the innovations we need in the public-service sector on new institutions, as historically we have always done?

The answer is that public-service institutions have become too important in developed countries, and too big. The public-service sector, both the governmental one and the nongovernmental but not-for-profit one, has grown faster during this century than the private sector—maybe three to five times as fast. The growth has been especially fast since World War II.

To some extent, this growth has been excessive. Wherever public-service activities can be converted into profit-making enterprises, they should be so converted. This applies not only to the kind of municipal services the city of Lincoln, Nebraska, now "privatizes." The move from non-profit to profit has already gone very far in the American hospital. I expect it to become a stampede in professional and graduate education. To subsidize the highest earners in developed society, the holders of advanced professional degrees, can hardly be justified.

A central economic problem of developed societies during the next twenty or thirty years is surely going to be capital formation; only in Japan is it still adequate for the economy's needs. We therefore can ill afford to have activities conducted as "non-profit," that is, as activities that devour capital rather than form it, if they can be organized as activities that form capital, as activities that make a profit.

But still the great bulk of the activities that are being discharged in and by public-service institutions will remain public-service activities, and will neither disappear nor be transformed. Consequently, they have to be made producing and productive. Public-service institutions

will have to learn to be innovators, to manage themselves entrepreneurially. To achieve this, public-service institutions will have to
learn to look upon social, technological, economic, and demographic
shifts as opportunities in a period of rapid change in all these areas.
Otherwise, they will become obstacles. The public-service institutions
will increasingly become unable to discharge their mission as they ad-
here to programs and projects that cannot work in a changed environment, and yet they will not be able or willing to abandon the missions
they can no longer discharge. Increasingly, they will come to look the
way the feudal barons came to look after they had lost all social function
around 1300: as parasites, functionless, with nothing left but the power
to obstruct and to exploit. They will become self-righteous while increasingly losing their legitimacy. Clearly, this is already happening to
the apparently most powerful among them, the labor union. Yet a
society in rapid change, with new challenges, new requirements and
opportunities, needs public-service institutions.

The public school in the United States exemplifies both the opportunity and the dangers. Unless it takes the lead in innovation it is unlikely
to survive this century, except as a school for the minorities in the slums.
For the first time in its history, the United States faces the threat of a
class structure in education in which all but the very poor remain
outside of the public school system—at least in the cities and suburbs
where most of the population lives. And this will squarely be the fault
of the public school itself because what is needed to reform the public
school is already known (see Chapter 9).

Many other public-service institutions face a similar situation. The
knowledge is there. The need to innovate is clear. They now have to
learn how to build entrepreneurship and innovation into their own
system. Otherwise, they will find themselves superseded by outsiders
who will create competing entrepreneurial public-service institutions
and so render the existing ones obsolete.

The late nineteenth century and early twentieth century was a
period of tremendous creativity and innovation in the public-service
field. Social innovation during the seventy-five years until the 1930s was
surely as much alive, as productive, and as rapid as technological innovation if not more so. But in these periods the innovation took the form
of creating new public-service institutions. Most of the ones we have
around now go back no more than sixty or seventy years in their present
form and with their present mission. The next twenty or thirty years

will be very different. The need for social innovation may be even greater, but it will very largely have to be social innovation within the existing public-service institution. To build entrepreneurial management into the existing public-service institution may thus be the foremost political task of this generation.

15

The New Venture

For the existing enterprise, whether business or public-service institution, the controlling word in the term "entrepreneurial management" is "entrepreneurial." For the new venture, it is "management." In the existing business, it is the existing that is the main obstacle to entrepreneurship. In the new venture, it is its absence.

The new venture has an idea. It may have a product or a service. It may even have sales, and sometimes quite a substantial volume of them. It surely has costs. And it may have revenues and even profits. What it does not have is a "business," a viable, operating, organized "present" in which people know where they are going, what they are supposed to do, and what the results are or should be. But unless a new venture develops into a new business and makes sure of being "managed," it will not survive no matter how brilliant the entrepreneurial idea, how much money it attracts, how good its products, nor even how great the demand for them.

Refusal to accept these facts destroyed every single venture started by the nineteenth century's greatest inventor, Thomas Edison. Edison's ambition was to be a successful businessman and the head of a big company. He should have succeeded, for he was a superb business planner. He knew exactly how an electric power company had to be set up to exploit his invention of the light bulb. He knew exactly how to get all the money he could possibly need for his ventures. His products were immediate successes and the demand for them practically insatiable. But Edison remained an entrepreneur; or rather, he thought that "managing" meant being the boss. He refused to build a management team. And so every one of his four or five companies collapsed ignominiously once it got to middle size, and was saved only by booting Edison himself out and replacing him with professional management.

Entrepreneurial management in the new venture has four requirements:

It requires, first, a focus on the market.

It requires, second, financial foresight, and especially planning for
cash flow and capital needs ahead.

It requires, third, building a top management team long before the
new venture actually needs one and long before it can actually afford
one.

And finally, it requires of the founding entrepreneur a decision in
respect to his or her own role, area of work, and relationships.

I

THE NEED FOR MARKET FOCUS

A common explanation for the failure of a new venture to live up
to its promise or even to survive at all is: "We were doing fine until these
other people came and took our market away from us. We don't really
understand it. What they offered wasn't so very different from what we
had." Or one hears: "We were doing all right, but these other people
started selling to customers we'd never even heard of and all of a
sudden they had the market."

When a new venture does succeed, more often than not it is in a
market other than the one it was originally intended to serve, with
products or services not quite those with which it had set out, bought
in large part by customers it did not even think of when it started, and
used for a host of purposes besides the ones for which the products were
first designed. If a new venture does not anticipate this, organizing itself
to take advantage of the unexpected and unseen markets; if it is not
totally market-focused, if not market-driven, then it will succeed only
in creating an opportunity for a competitor.

There are exceptions, to be sure. A product designed for one specific
use, especially if scientific or technical, often stays with the market and
the end use for which it was designed. But not always. Even a prescription drug designed for a specific ailment and tested for it sometimes
ends up being used for some other quite different ailment. One example is a compound that is effectively used in the treatment of stomach
ulcers. Or a drug designed primarily for the treatment of human beings
may find its major market in veterinary medicine.

Anything genuinely new creates markets that nobody before even imagined. No one knew that he needed an office copier before the first Xerox machine came out around 1960; five years later no business could imagine doing without a copier. When the first jet planes started to fly, the best market research pointed out that there were not even enough passengers for all the transatlantic liners then in service or being built. Five years later the transatlantic jets were carrying fifty to one hundred times as many passengers each year as had ever before crossed the Atlantic.

The innovator has limited vision, in fact, he has tunnel-vision. He sees the area with which he is familiar—to the exclusion of all other areas.

An example is DDT. Designed during World War II to protect American soldiers against tropical insects and parasites, it eventually found its greatest application in agriculture to protect livestock and crops against insects—to the point where it had to be banned for being too effective. Yet not one of the distinguished scientists who designed DDT during World War II envisaged these uses of DDT. Of course they knew that babies die from fly-borne "summer" diarrhea. Of course they knew that livestock and crops are infested by insect parasites. But these things they knew as laymen. As experts, they were concerned with the tropical diseases of humans. It was the ordinary American soldier who then applied DDT to the areas in which he was the "expert," that is, to his home, his cows, his cotton patch.

Similarly, the 3M Company did not see that an adhesive tape it had developed for industry would find myriad uses in the household and in the office—becoming Scotch Tape. 3M had for many years been a supplier of abrasives and adhesives to industry, and moderately successful in industrial markets. It had never even thought of consumer markets. It was pure accident which led the engineer who had designed an industrial product no industrial user wanted to the realization that the stuff might be salable in the consumer market. As the story goes, he took some samples home when the company had already decided to abandon the product. To his surprise, his teenage daughters began to use it to hold their curls overnight. The only unusual thing about this story is that he and his bosses at 3M recognized that they had stumbled upon a new market.

A German chemist developed Novocain as the first local anesthetic in 1905. But he could not get the doctors to use it; they preferred total anesthesia (they only accepted Novocain during World War I). But totally unexpectedly, dentists began to use the stuff. Whereupon—or so the story goes—the chemist began to travel up and down Germany making speeches against Novocain's use in dentistry. He had not designed it for that purpose!

That reaction was somewhat extreme, I admit. Still, entrepreneurs *know* what their innovation is meant to do. And if some other use for it appears, they tend to resent it. They may not actually refuse to serve customers they have not "planned" for, but they are likely to make it clear that these customers are not welcome.

This is what happened with the computer. The company that had the first computer, Univac, knew that its magnificent machine was designed for scientific work. And so it did not even send a salesman out when a business showed interest in it; surely, it argued, these people could not possibly know what a computer was all about. IBM was equally convinced that the computer was an instrument for scientific work: their own computer had been designed specifically for astronomical calculations. But IBM was willing to take orders from businesses and to serve them. Ten years later, around 1960, Univac still had by far the most advanced and best machine. IBM had the computer market.

The textbook prescription for this problem is "market research." But it is the wrong prescription.

One cannot do market research for something genuinely new. One cannot do market research for something that is not yet on the market. Around 1950, Univac's market research concluded that, by the year 2000, about one thousand computers would be sold; the actual figure in 1984 was about one million. And yet this was the most "scientific," careful, rigorous market research ever done. There was only one thing wrong with it: it started out with the assumption, then shared by everyone, that computers were going to be used for advanced scientific work —and for that use, the number is indeed quite limited. Similarly, several companies who turned down the Xerox patents did so on the basis of thorough market research which showed that printers had no use at all for a copier. Nobody had any inkling that businesses, schools, universities, colleges, and a host of private individuals would want to buy a copier.

The new venture therefore needs to start out with the assumption that its product or service may find customers in markets no one thought of, for uses no one envisaged when the product or service was designed, and that it will be bought by customers outside its field of vision and even unknown to the new venture.

If the new venture does not have such a market focus from the very beginning, all it is likely to create is the market for a competitor. A few years later "those people" will come in and take away "our market," or "those other people" who started "selling to customers we'd never even heard of" all of a sudden will indeed have preempted the market.

To build market focus into a new venture is not in fact particularly difficult. But what is required runs counter to the inclinations of the typical entrepreneur. It requires, first, that the new venture systematically hunt out both the unexpected success and the unexpected failure (cf. Chapter 3). Rather than dismiss the unexpected as an "exception," as entrepreneurs are inclined to do, they need to go out and look at it carefully and as a distinct opportunity.

Shortly after World War II, a small Indian engineering firm bought the license to produce a European-designed bicycle with an auxiliary light engine. It looked like an ideal product for India; yet it never did well. The owner of this small firm noticed, however, that substantial orders came in for the engines alone. At first he wanted to turn down those orders; what could anyone possibly do with such a small engine? It was curiosity alone that made him go to the actual area the orders came from. There he found farmers were taking the engines off the bicycles and using them to power irrigation pumps that hitherto had been hand-operated. This manufacturer is now the world's largest maker of small irrigation pumps, selling them by the millions. His pumps have revolutionized farming all over Southeast Asia.

To be market-driven also requires that the new venture be willing to experiment. If there is any interest in the new venture's product or service on the part of consumers or markets that were not in the original plan, one tries to find somebody in that new and unexpected area who might be willing to test the new product or service and find out what, if any, application it might have. One provides free samples to people in the "improbable" market to see what they can do with it, whether they can use the stuff at all, or what it would have to be like for them to become customers for it. One advertises in the trade papers of the industry whence indications of interest came, and so on.

The DuPont Company never thought of automobile tires as a major application for the new Nylon fiber it had developed. But when one of the Akron tire manufacturers showed interest in trying out Nylon,

DuPont set up a plant. A few years later, tires had become Nylon's biggest and most profitable market.

It does not require a great deal of money to find out whether an unexpected interest from an unexpected market is an indication of genuine potential or a fluke. It requires sensitivity and a little systematic work.

Above all, the people who are running a new venture need to spend time outside: in the marketplace, with customers and with their own salesmen, looking and listening. The new venture needs to build in systematic practices to remind itself that a "product" or a "service" is defined by the customer, not by the producer. It needs to work continuously on challenging itself in respect to the utility and value that its products or services contribute to customers.

The greatest danger for the new venture is to "know better" than the customer what the product or service is or should be, how it should be bought, and what it should be used for. Above all, the new venture needs willingness to see the unexpected success as an opportunity rather than as an affront to its expertise. And it needs to accept that elementary axiom of marketing: Businesses are not paid to reform customers. They are paid to satisfy customers.

II

FINANCIAL FORESIGHT

Lack of market focus is typically a disease of the "neo-natal," the infant new venture. It is the most serious affliction of the new venture in its early stages—and one that can permanently stunt even those that survive.

The lack of adequate financial focus and of the right financial policies is, by contrast, the greatest threat to the new venture in the next stage of its growth. It is, above all, a threat to the rapidly growing new venture. The more successful a new venture is, the more dangerous the lack of financial foresight.

Suppose that a new venture has successfully launched its product or service and is growing fast. It reports "rapidly increasing profits" and issues rosy forecasts. The stock market then "discovers" the new venture, especially if it is high-tech or in a field otherwise currently fashionable. Predictions abound that the new venture's sales will reach a billion dollars within five years. Eighteen months later, the new venture collapses. It may not go out of existence or go bankrupt. But it is suddenly awash in red ink, lays off 180 of its 275 employees, fires the president, or is sold at a bargain price to a big company. The causes are always the same: lack of cash; inability to raise the capital needed for expansion; and loss of control, with expenses, inventories, and receivables in disarray. These three financial afflictions often hit together at the same time. Yet any one of them by itself endangers the health, if not the life, of the new venture.

Once this financial crisis has erupted, it can be cured only with great difficulty and considerable suffering. But it is eminently preventable.

Entrepreneurs starting new ventures are rarely unmindful of money; on the contrary, they tend to be greedy. They therefore focus on profits. But this is the wrong focus for a new venture, or rather, it comes last rather than first. Cash flow, capital, and controls come much earlier. Without them, the profit figures are fiction—good for twelve to eighteen months, perhaps, after which they evaporate.

Growth has to be fed. In financial terms this means that growth in a new venture demands adding financial resources rather than taking them out. Growth needs more cash and more capital. If the growing new venture shows a "profit" it is a fiction: a bookkeeping entry put in only to balance the accounts. And since taxes are payable on this fiction in most countries, it creates a liability and a cash drain rather than "surplus." The healthier a new venture and the faster it grows, the more financial feeding it requires. The new ventures that are the darlings of the newspapers and the stock market letters, the new ventures that show rapid profit growth and "record profits," are those most likely to run into desperate trouble a couple of years later.

The new venture needs cash flow analysis, cash flow forecasts, and cash management. The fact that America's new ventures of the last few years (with the significant exception of high-tech companies) have been doing so much better than new ventures used to do is largely because the new entrepreneurs in the United States have learned that entrepreneurship demands financial management.

Cash management is fairly easy if there are reliable cash flow forecasts, with "reliable" meaning "worst case" assumptions rather than hopes. There is an old banker's rule of thumb, according to which in forecasting cash income and cash outlays one assumes that bills will have to be paid sixty days earlier than expected and receivables will come in sixty days later. If the forecast is overly conservative, the worst that can happen—it rarely does in a growing new venture—is a temporary cash surplus.

A growing new venture should know twelve months ahead of time how much cash it will need, when, and for what purposes. With a year's lead time, it is almost always possible to finance cash needs. But even if a new venture is doing well, raising cash in a hurry and in a "crisis" is never easy and always prohibitively expensive. Above all, it always sidetracks the key people in the company at the most critical time. For several months they then spend their time and energy running from one financial institution to another and cranking out one set of questionable financial projections after another. In the end, they usually have to mortgage the long-range future of the business to get through a ninety-day cash bind. When they finally are able again to devote time and thought to the business, they have irrevocably missed the major opportunities. For the new venture, almost by definition, is under cash pressure when the opportunities are greatest.

The successful new venture will also outgrow its capital structure. A rule of thumb with a good deal of empirical evidence to support it says that a new venture outgrows its capital base with every increase in sales (or billings) of the order of 40 to 50 percent. After such growth, a new venture also needs a new and different capital structure, as a rule. As the venture grows, private sources of funds, whether from the owners and their families or from outsiders, become inadequate. The company has to find access to much larger pools of money by going "public," by finding a partner or partners among established companies, or by raising money from insurance companies and pension funds. A new venture that had been financed by equity money now needs to shift to long-term debt, or vice versa. As the venture grows, the existing capital structure always becomes the wrong structure and an obstacle.

In some new ventures, capital planning is comparatively easy. When the business consists of uniform and entirely local units—restaurants in a chain, freestanding surgical centers or individual hospitals in different cities, homebuilders with separate operations in a number of different metropolitan areas, specialty stores and the like—each unit can be financed as a separate business. One solution is franchising (which is, in essence, a way to finance rapid expansion). Another is setting up each local unit as a company, with separate and often local investors as "limited" partners. The capital needed for growth and expansion can thus be raised step by step, and the success of the preceding unit furnishes documentation and the incentive for the investors in the succeeding ones. But it only works when: (a) each unit breaks even fairly soon, at most perhaps within two or three years; (b) when the operation can be made routine, so that people of limited managerial competence —the typical franchise holder, or the business manager of a local free-standing surgical center—can do a decent job without much supervision; and (c) when the individual unit itself reaches fairly swiftly the optimum size beyond which it does not require further capital but produces cash surplus to help finance the startup of additional units.

For new ventures other than those capable of being financed as separate units, capital planning is a survival necessity. If a growing new venture plans realistically—and that again means assuming the maximum rather than the minimum need—for its capital requirement and its capital structure three years ahead, it should normally have little difficulty in obtaining the kind of money it needs, when it needs it, and in the form in which it needs it. If it waits until it outgrows its capital base and its capital structure, it is putting its survival—and most assuredly its independence—on the block. At the very least, the founders will find that they have taken all the entrepreneurial risk and worked hard only to make other people the rich owners. From being owners, they will have become employees, with the new investors taking control.

Finally, the new venture needs to plan the financial system it requires to manage growth. Again and again, a growing new venture starts off with an excellent product, excellent standing in its market, and excellent growth prospects. Then suddenly everything goes out of control: receivables, inventory, manufacturing costs, administrative costs, service, distribution, everything. Once one area gets out of control, all of them do. The enterprise has outgrown its control structure. By the time control has been reestablished, markets have been lost, customers have become disgruntled if not hostile, distributors have lost their confidence in the company. Worst of all, employees have lost trust in management, and with good reason.

Fast growth always makes obsolete the existing controls. Again, a growth of 40 to 50 percent in volume seems to be the critical figure. Once control has been lost, it is hard to recapture. Yet the loss of control can be prevented quite easily. What is needed is first to think through the critical areas in a given enterprise. In one, it may be product quality; in another, service; in a third, receivables and inventory; in a fourth, manufacturing costs. Rarely are there more than four or five critical areas in any given enterprise. (Managerial and administrative overhead should, however, always be included. A disproportionate and fast increase in the percentage of revenues absorbed by managerial and administrative overhead, which means that the enterprise hires managerial and administrative people faster than it actually grows, is usually the first sign that a business is getting out of control, that its management structure and practices are no longer adequate to the task.)

To live up to its growth expectations, a new venture must establish today the controls in these critical areas it will need three years hence. Elaborate controls are not necessary nor does it matter that the figures are only approximate. What matters is that the management of the new venture is aware of these critical areas, is being reminded of them, and can thus act fast if the need arises. Disarray normally does not appear if there is adequate attention to the key areas. Then the new venture will have the controls it needs when it needs them.

Financial foresight does not require a great deal of time. It does require a good deal of thought, however. The technical tools to do the job are easily available; they are spelled out in most texts on managerial accounting. But the work will have to be done by the enterprise itself.

III

BUILDING A TOP MANAGEMENT TEAM

The new venture has successfully established itself in the right market and has then successfully found the financial structure and the financial system it needs. Nonetheless, a few years later it is still prone to run into a serious crisis. Just when it appears to be on the threshold of becoming an "adult"—a successful, established, going concern—it gets into trouble nobody seems to understand. The products are first-rate, the prospects are excellent, and yet the business simply cannot grow. Neither profitability nor quality, nor any of the other major areas performs.

The reason is always the same: a lack of top management. The business has outgrown being managed by one person, or even two people, and it now needs a management team at the top. If it does not have one already in place at the time, it is very late—in fact, usually too late. The best one can then hope is that the business will survive. But it is likely to be permanently crippled or to suffer scars that will bleed for many years to come. Morale has been shattered and employees throughout the company are disillusioned and cynical. And the people who founded the business and built it almost always end up on the outside, embittered and disenchanted.

The remedy is simple: To build a top management team *before* the venture reaches the point where it must have one. Teams cannot be formed overnight. They require long periods before they can function. Teams are based on mutual trust and mutual understanding, and this takes years to build up. In my experience, three years is about the minimum.

But the small and growing new venture cannot afford a top management team; it cannot sustain half a dozen people with big titles and corresponding salaries. In fact, in the small and growing business, a very small number of people do everything as it comes along. How, then, can one square this circle?

Again, the remedy is relatively simple. But it does require the will on the part of the founders to build a team rather than to keep on running everything themselves. If one or two people at the top believe that they, and they alone, must do everything, then a management crisis a few months, or at the latest, a few years down the road becomes inevitable.

Whenever the objective economic indicators of a new venture—market surveys, for instance, or demographic analysis—indicate that the business may double within three or five years, then it is the duty of the founder or founders to build the management team the new venture will very soon require. This is preventive medicine, so to speak.

First of all the founders, together with other key people in the firm, will have to think through the key activities of their business. What are the specific areas upon which the survival and success of this particular business depend? Most of the areas will be on everyone's list. But if there are divergencies and dissents—and there should be on a question as important as this—they should be taken seriously. Every activity which any member of the group thinks belongs there should go down on the list.

The key activities are not to be found in books. They emerge from analysis of the specific enterprise. Two enterprises that to an outsider appear to be in an identical line of business may well end up defining their key activities quite differently. One, for instance, may put production in the center; the other, customer service. Only two key activities are always present in any organization: there is always the management of people and there is always the management of money. The rest has to be determined by the people within looking at the enterprise and at their own jobs, values, and goals.

The next step is, then, for each member of the group, beginning with the founder, to ask: "What are the activities that *I* am doing well? And what are the activities that each of my key associates in this business is actually doing well?" Again, there is going to be agreement on most of the people and on most of their strengths. But, again, any disagreement should be taken seriously.

Next, one asks: "Which of the key activities should each of us, therefore, take on as his or her first and major responsibility because they fit the individual's strengths? Which individual fits which key activity?"

Then the work on building a team can begin. The founder starts to discipline himself (or herself) not to handle people and their problems, if this is not the key activity that fits him best. Perhaps this individual's key strength is new products and new technology. Perhaps this individual's key activity is operations, manufacturing, physical distribution, service. Or perhaps it is money and finance and someone else had better handle people. But all key activities need to be covered by someone who has proven ability in performance.

There is no rule that says "A chief executive has to be in charge of this or that." Of course a chief executive is the court of last resort and has ultimate accountability. And the chief executive also has to make sure of getting the information necessary to discharge this ultimate accountability. The chief executive's own *work*, however, depends on what the enterprise requires and on who the individual is. As long as the CEO's work program consists of key activities, he or she does a CEO's job. But the CEO also is responsible for making sure that all the other key activities are adequately covered.

Finally, goals and objectives for each area need to be set. Everyone who takes on the primary responsibility for a key activity, whether product development or people, or money, must be asked: "What can this enterprise expect of *you?* What should we hold *you* accountable for? What are *you* trying to accomplish and by what time?" But this is elementary management, of course.

It is prudent to establish the top management team informally at first. There is no need to give people titles in a new and growing venture, nor to make announcements, nor even to pay extra. All this can wait a year or so, until it is clear that the new setup works, and how. In the meantime, all the members of the team have much to learn: their job; how they work together; and what they have to do to enable the CEO and their colleagues to do their jobs. Two or three years later, when the growing venture needs a top management, it has one.

However, should it fail to provide for a top management before it actually needs one, it will lose the capacity to manage itself long before it actually needs a top management team. The founder will have become so overloaded that important tasks will not get done. At this point the company can go one of two ways. The first possibility is that the founder concentrates on the one or two areas that fit his or her abilities and interests. These are key areas indeed, but they are not the only crucial ones, and no one is then left to look after the others. Two years later, important areas have been slighted and the business is in dire straits. The other, worse, possibility is that the founder is conscientious. He knows that people and money are key activities and need to be taken care of. His own abilities and interests, which actually built the business, are in the design and development of new products. But being conscientious, the founder forces himself to take care of people and finance. Since he is not very gifted in either area, he does poorly in both. It also takes him forever to reach decisions or to do any work in these areas, so that he is forced, by lack of time, to neglect what he is really good at and what the company depends on him for, the development of new technology and new products. Three years later the company will have become an empty shell without the products it needs, but also without the management of people and the management of money it needs.

In the first example, it may be possible to save the company. After all, it has the products. But the founder will inevitably be removed by whoever comes in to salvage the company. In the second case, the company usually cannot be saved at all and has to be sold or liquidated.

Long before it has reached the point where it needs the balance of a top management team, the new venture has to create one. Long before the time has come at which management by one person no longer works and becomes mismanagement, that one person also has to start learning how to work with colleagues, has to learn to trust people, yet also how to hold them accountable. The founder has to learn to become the leader of a team rather than a "star" with "helpers."

I V

"WHERE CAN I CONTRIBUTE?"

Building a top management team may be the single most important step toward entrepreneurial management in the new venture. It is only the first step, however, for the founders themselves, who then have to think through what their own future is to be.

As a new venture develops and grows, the roles and relationships of the original entrepreneurs inexorably change. If the founders refuse to accept this, they will stunt the business and may even destroy it.

Every founder-entrepreneur nods to this and says, "Amen." Everyone has horror stories of other founder-entrepreneurs who did not change as the venture changed, and who then destroyed both the business and themselves. But even among the founders who can accept that they themselves need to do something, few know how to tackle changing their own roles and relationships. They tend to begin by asking: "What do I like to do?" Or at best, "Where do I fit in?" The right question to start with is: "What will the venture need *objectively* by way of management from here on out?" And in a growing new venture, the founder has to ask this question whenever the business (or the public-service institution) grows significantly or changes direction or character, that is, changes its products, services, markets, or the kind of people it needs.

The next question the founder must ask is: "What am I good at? What, of all these needs of the venture, could I supply, and supply with distinction?" Only after having thought through these two questions should a founder then ask: "What do I really want to do, and believe in doing? What am I willing to spend years on, if not the rest of my life? Is this something the venture really needs? Is it a major, essential, indispensable contribution?"

One example is that of the successful American post–World War II metropolitan university, Pace, in New York City. Dr. Edward Mortola built up the institution from nothing in 1947 into New York City's third-largest and fastest-growing university, with 25,000 students and well-regarded graduate schools. In the university's early years he was a radical innovator. But when Pace was still very small (around 1950), Mortola built a strong top management team. All members were given a major, clearly defined responsibility, for which they were expected to take full accountability and give leadership. A few years later, Mortola then decided what his own role was to be and converted himself into a traditional university president, while at the same time building a strong independent board of trustees to advise and support him.

But the questions of what a venture needs, what the strengths of the founder-entrepreneur are, and what he or she wants to do, might be answered quite differently.

Edwin Land, for instance, the man who invented Polaroid glass and the Polaroid camera, ran the company during the first twelve or fifteen years of its life, until the early 1950s. Then it began to grow fast. Land thereupon designed a top management team and put it in place. As for himself, he decided that he was not the right man for the top management job in the company: what he and he alone could contribute was scientific innovation. Accordingly, Land built himself a laboratory and established himself as the company's consulting director for basic research. The company itself, in its day-to-day operations, he left to others to run.

Ray Kroc, the man who conceived and built McDonald's, reached a similar conclusion. He remained president until he died well past age eighty. But he put a top management team in place to run the company and appointed himself the company's "marketing conscience." Until shortly before his death, he visited two or three McDonald's restaurants each week, checking their quality carefully, the level of cleanliness and friendliness. Above all, he looked at the customers, talked to them and listened to them. This enabled the company to make the necessary changes to retain its leadership in the fast-food industry.

Similarly, in a much smaller new venture, a building supply company in the Pacific Northwest of the United States, the young man who built the company decided that his role was not to run the company but to develop its critical resource, the managers who are responsible for its two hundred branches in small towns and suburbs. These managers are in effect running their own local business. They are supported by strong services in headquarters: central buying, quality control, control of credit and receivables, and so on. But the selling is done by each manager, locally and with very little help—maybe one salesman and a couple of truck drivers.

The business depends on the motivation, drive, ability, and enthusiasm of these isolated, fairly unsophisticated individuals. None of them has a college degree and few have even finished high school. So the founder of this company makes it his business to spend twelve to fifteen days each month in the field visiting branch managers, spending half a day with them, discussing their business, their plans, their aspirations. This may well be the only distinction the company has—otherwise, every other building materials wholesaler does the same things. But this performance of the one key activity by the chief executive has enabled the company to grow three to four times as fast as any competitor, even in recession times.

Yet another quite different answer to the same question was given by the three scientists who, together, founded what has become one of the largest and most successful companies in the semiconductor industry. When they asked themselves, "What are the needs of the business?" the answer was that there were three: "One for basic business strategy, one for scientific research and development, and one for the development of people—especially scientific and technical people." They decided which of the three was most suited for each of these assignments, and then divided them according to their strengths. The person who took the human relations and human development job had actually been a prolific scientific innovator and had high standing in scientific circles. But he decided, and his colleagues concurred, that he was superbly fitted for the managerial, the people task, so he took it. "It was not," he once said in a speech, "what I really wanted to do, but it was where I could make the greatest contribution."

These questions may not always lead to such happy endings. They may even lead to the decision to leave the company.

In one of the most successful new financial services ventures in the United States, this is what the founder concluded. He did establish a top management team. He asked what the company needed. He looked at himself and his strengths; and he found no match between the needs of the company and his own abilities, let alone between the needs of the company and the things he wanted to do. "I trained my own successor for about eighteen months, then turned the company over to him and resigned," he said. Since then he has started three new businesses, not one of them in finance, has developed them successfully to medium size, and then quit again. He wants to develop new businesses but does not enjoy running them. He accepts that both the businesses and he are better off divorced from one another.

Other entrepreneurs in this same situation might reach different conclusions. The founder of a well-known medical clinic, a leader in its particular field, faced a similar dilemma. The needs of the institution were for an administrator and money-raiser. His own inclinations were to be a researcher and a clinician. But he realized that he was good at raising money and capable of learning to be the chief executive officer of a fairly large health-care organization. "And so," he says, "I felt it my duty to the venture I had created, and to my associates in it, to suppress my own desires and to take on the job of chief administrator and money-raiser. But I would never have done so had I not known that I had the abilities to do the job, and if my advisors and my board had not all assured me that I had these abilities."

The question, "Where do *I* belong?" needs to be faced up to and thought through by the founder-entrepreneur as soon as the venture shows the first signs of success. But the question can be faced up to much earlier. Indeed, it might be best thought through before the new venture is even started.

This is what Soichiro Honda, the founder and builder of Honda Motor Company in Japan, did when he decided to open a small business in the darkest days after Japan's defeat in World War II. He did not start his venture until he had found the right man to be his partner and to run administration, finance, distribution, marketing, sales, and personnel. For Honda had decided from the outset that he belonged in engineering and production and would not run anything else. This decision made the Honda Motor Company.

There is an earlier and even more instructive example, that of Henry Ford. When Ford decided in 1903 to go into business for himself, he did exactly what Honda did forty years later: before starting, he found the right man to be his partner and to run the areas where Ford knew he did not belong—administration, finance, distribution, marketing, sales, and personnel. Like Honda, Henry Ford knew that he belonged in engineering and manufacturing and was going to confine himself to these two areas. The man he found, James Couzens,* contributed as much as Ford to the success of the company. Many of the

---

*Who later became mayor of Detroit and senator from Michigan, and might well have become President of the United States had he not been born in Canada.

best known policies and practices of the Ford Motor Company for which Henry Ford is often given credit—the famous $5-a-day wage of 1913, or the pioneering distribution and service policies, for example —were Couzens's ideas and at first resisted by Ford. So effective did Couzens become that Ford grew increasingly jealous of him and forced him out in 1917. The last straw was Couzens's insistence that the Model T was obsolescent and his proposal to use some of the huge profits of the company to start work on a successor.

The Ford Motor Company grew and prospered to the very day of Couzens's resignation. Within a few short months thereafter, as soon as Henry Ford had taken every single top management function into his own hands, forgetting that he had known earlier where he belonged, the Ford Motor Company began its long decline. Henry Ford clung to the Model T for a full ten years, until it had become literally unsalable. And the company's decline was not reversed for thirty years after Couzens's dismissal until, with his grandfather dying, a very young Henry Ford II took over the practically bankrupt business.

THE NEED FOR OUTSIDE ADVICE

These last cases point up an important factor for the entrepreneur in the new and growing venture, the need for independent, objective outside advice.

The growing new venture may not need a formal board of directors. Moreover, the typical board of directors very often does not provide the advice and counsel the founder needs. But the founder does need people with whom he can discuss basic decisions and to whom he listens. Such people are rarely to be found within the enterprise. Somebody has to challenge the founder's appraisal of the needs of the venture, and of his own personal strengths. Someone who is not a part of the problem has to ask questions, to review decisions and, above all, to push constantly to have the long-term survival needs of the new venture satisfied by building in the market focus, supplying financial foresight, and creating a functioning top management team. This is the final requirement of entrepreneurial management in the new venture.

The new venture that builds such entrepreneurial management into its policies and practices will become a flourishing large business.*

*A fine description of this process is to be found in *High-Output Management* (New

In so many new ventures, especially high-tech ventures, the techniques discussed in this chapter are spurned and even despised. The argument is that they constitute "management" and "We are entrepreneurs." But this is not informality; it is irresponsibility. It confuses manners and substance. It is old wisdom that there is no freedom except under the law. Freedom without law is license, which soon degenerates into anarchy, and shortly thereafter into tyranny. It is precisely because the new venture has to maintain and strengthen the entrepreneurial spirit that it needs foresight and discipline. It needs to prepare itself for the demands its own success will make of it. Above all, it needs responsibility—and this, in the last analysis, is what entrepreneurial management supplies to the new venture.

There is much more that could be said about managing the new venture, about financing, staffing, marketing its products, and so on. But these specifics are adequately covered in a number of publications.* What this chapter has tried to do is to identify and discuss the few fairly simple policies that are crucial to the survival and success of any new venture, whether a business or a public-service institution, whether "high-tech," "low-tech," or "no-tech," whether started by one man or woman or by a group, and whether intended to remain a small business or to become "another IBM."

York: Random House, 1983), by Andrew S. Grove, co-founder and president of Intel, one of the largest manufacturers of semiconductors.

*For some of these, see the Suggested Readings at the back of this book.

III

ENTREPRENEURIAL
STRATEGIES

—

Just as entrepreneurship requires entrepreneurial management, that is, practices and policies within the enterprise, so it requires practices and policies outside, in the marketplace. It requires entrepreneurial strategies.

16

"Fustest with the Mostest"

Of late, "strategy in business"* has become the "in" word, with any number of books written about it.† However, I have not come across any discussion of entrepreneurial strategies. Yet they are important; they are distinct; and they are different.

There are four specifically entrepreneurial strategies:

1. Being "Fustest with the Mostest";
2. "Hitting Them Where They Ain't";
3. Finding and occupying a specialized "ecological niche";
4. Changing the economic characteristics of a product, a market, or an industry.

These four strategies are not mutually exclusive. One and the same entrepreneur often combines two, sometimes even elements of three, in one strategy. They are also not always sharply differentiated; the same strategy might, for instance, be classified as "Hitting Them Where They Ain't" or as "Finding and occupying a specialized 'ecological niche.' " Still, each of these four has its prerequisites. Each fits certain kinds of innovation and does not fit others. Each requires specific behavior on the part of the entrepreneur. Finally, each has its own limitations and carries its own risks.

*The 1952 edition of the *Concise Oxford Dictionary* still defined strategy as: "Generalship; the art of war; management of an army or armies in a campaign." Alfred D. Chandler, Jr., first applied the term to the conduct of a business in 1962 in his pioneering *Strategy and Structure* (Cambridge, Mass.: M.I.T. Press), which studied the evolution of management in the big corporation. But shortly thereafter, in 1963, when I wrote the first analysis of business strategy, the publisher and I found that the word could not be used in the title without risk of serious misunderstanding. Booksellers, magazine editors, and senior business executives all assured us that "strategy" for them meant the conduct of military or election campaigns. The book discussed most that is now considered "strategy." It uses the word in the text. But the title we chose was *Managing for Results*.
†Of which I have found Michael Porter's *Competitive Strategies* (New York: Free Press, 1980) the most useful.

I

BEING "FUSTEST WITH THE MOSTEST"

Being "Fustest with the Mostest" was how a Confederate cavalry general in America's Civil War explained consistently winning his battles. In this strategy the entrepreneur aims at leadership, if not at dominance of a new market or a new industry. Being "Fustest with the Mostest" does not necessarily aim at creating a big business right away, though often this is indeed the aim. But it aims from the start at a permanent leadership position.

Being "Fustest with the Mostest" is the approach that many people consider the entrepreneurial strategy *par excellence*. Indeed, if one were to go by the popular books on entrepreneurs,* one would conclude that being "Fustest with the Mostest" is the only entrepreneurial strategy—and a good many entrepreneurs, especially the high-tech ones, seem to be of the same opinion.

They are wrong, however. To be sure, a good many entrepreneurs have indeed chosen this strategy. Yet being "Fustest with the Mostest" is not even the dominant entrepreneurial strategy, let alone the one with the lowest risk or the highest success ratio. On the contrary, of all entrepreneurial strategies it is the greatest gamble. And it is unforgiving, making no allowances for mistakes and permitting no second chance.

But if successful, being "Fustest with the Mostest" is highly rewarding.

Here are some examples to show what this strategy consists of and what it requires.

Hoffmann-LaRoche of Basel, Switzerland, has for many years been the world's largest and in all probability its most profitable pharmaceutical company. But its origins were quite humble: until the mid-1920s, Hoffmann-LaRoche was a small and struggling manufacturing chemist, making a few textile dyes. It was totally overshadowed by the huge German dye-stuff makers and two or three much bigger chemical firms in its own country. Then it gambled on the newly discovered vitamins at a time when the scientific world still could not quite accept

*E.g., George Gilder's *The Spirit of Enterprise* (New York: Simon & Schuster, 1984), perhaps the most readable recent example of the genre.

that such substances existed. It acquired the vitamin patents—nobody else wanted them. It hired the discoverers away from Zürich University at several times the salaries they could hope to get as professors, salaries even industry had never paid before. And it invested all the money it had and all it could borrow in manufacturing and marketing these new substances.

Sixty years later, long after all vitamin patents have expired, Hoffmann-LaRoche has nearly half the world's vitamin market, now amounting to billions of dollars a year. The company followed the same strategy twice more: in the 1930s, when it went into the new sulfa drugs even though most scientists of the time "knew" that systemic drugs could not be effective against infections; and twenty years later, in the mid-fifties, when it went into the muscle-relaxing tranquilizers, Librium and Valium—at that time considered equally heretical and incompatible with what "every scientist knew."

DuPont followed the same strategy. When it came up with Nylon, the first truly synthetic fiber, after fifteen years of hard, frustrating research, DuPont at once mounted massive efforts, built huge plants, went into mass advertising—the company had never before had consumer products to advertise—and created the industry we now call plastics.

These are "big-company" stories, it will be said. But Hoffmann-LaRoche was not a big company when it started. And here are some more recent examples of companies that started from nothing with a strategy of getting there "Fustest with the Mostest."

The word processor is not much of a "scientific" invention. It hooks up three existing instruments: a typewriter, a display screen, and a fairly elementary computer. But this combination of existing elements has resulted in a genuine innovation that is radically changing office work. Dr. An Wang was a lone entrepreneur when he conceived of the combination some time in the mid-fifties. He had no track record as an entrepreneur and a minimum of financial backing. Yet he clearly aimed from the beginning at creating a new industry and at changing office work—and Wang Laboratories has, of course, become a very big company.

Similarly, the two young engineers who started the Apple computer in the proverbial garage, without financial backers or previous business experience, aimed from the beginning at creating an industry and dominating it.

Not every "Fustest with the Mostest" strategy needs to aim at creating a big business, though it must always aim at creating a business that dominates its market. The 3M Company in St. Paul, Minnesota, does not —as a matter of deliberate policy, it seems—attempt an innovation that might result in a big business by itself. Nor does Johnson & Johnson, the health-care and hygiene producer. Both companies are among the most fertile and most successful innovators. Both look for innovations that will lead to medium-sized rather than to giant enterprises, which are, however, dominant in their markets.

Being "Fustest with the Mostest" is not confined to businesses. It is also available to public-service institutions. When Wilhelm von Humboldt founded the University of Berlin in 1809—an event mentioned before in this book—he clearly aimed at being "Fustest with the Mostest." Prussia had just been defeated by Napoleon and had barely escaped total dismemberment. It was bankrupt, politically, militarily, and, above all, financially. It looked very much the way Germany was to look after Hitler's defeat in 1945. Yet Humboldt went out to build the largest university the Western world had ever seen or heard of— three to four times as large as anything then in existence. He went out to hire the leading scholars in every single discipline, beginning with the foremost philosopher of the time, Georg W. F. Hegel. And he paid his professors up to ten times as much as professors had ever been paid before, at a period when first-class scholars were going begging since the Napoleonic wars had forced many old and famous universities to disband.

A hundred years later, in the early years of this century, two surgeons in Rochester, an obscure Minnesota town far from population centers or medical schools, decided to establish a medical center based on totally new—and totally heretical—concepts of medical practice, and especially on building teams in which outstanding specialists would work together under a coordinating team leader. Frederick William Taylor, the so-called father of scientific management, had never met the Mayo Brothers. But in his well-known testimony before the Congress in 1911, he called the Mayo Clinic the "only complete and successful scientific management" he knew. These unknown provincial surgeons aimed from the beginning at dominance of the field, at attracting outstanding practitioners in every branch of medicine and the most gifted of the younger men, and at attracting also patients able and willing to pay what were then outrageous fees.

And twenty-five years later, the strategy of being "Fustest with the Mostest" was used by the March of Dimes to organize research into infantile paralysis (polio). Instead of aiming at gathering new knowledge step by step, as all earlier medical research had done, the March of Dimes aimed from the beginning at total victory over a completely mysterious disease. No one before had ever organized a "research lab without walls," in which a large number of scientists in a multitude of research institutions were commissioned to work on specific stages of a planned and managed research program. The March of Dimes established the pattern on which the United States, a little later, organized the first great research projects of World War II: the atom bomb, the radar lab, the proximity fuse, and then another fifteen years later, "Putting a Man on the Moon"—all innovative efforts using the "Fustest with the Mostest" strategy.

These examples show, first, that being "Fustest with the Mostest" requires an ambitious aim; otherwise it is bound to fail. It always aims at creating a new industry or a new market. At the least, as in the case of the Mayo Clinic or the March of Dimes, being "Fustest with the Mostest" aims at creating a quite different and highly unconventional process. The DuPonts surely did not say to themselves in the mid-twenties when they brought in Carothers: "We will establish the plastics industry" (indeed, the term was rarely used until the 1950s). But enough of the internal DuPont documents of the time have been published to show that the top management people did aim at creating a new industry. They were far from convinced that Carothers and his research would succeed. But they knew that they would have founded something big and brand new in the event of success, and something that would go far beyond a single product or even beyond a single major product line. Dr. Wang did not coin the term "the Office of the Future," as far as I know. But in his first advertisements, he announced a new office environment and new concepts of office work. Both the DuPonts and Wang from the beginning clearly aimed at dominating the industry they hoped they would succeed in creating.

The best example of what is implied in the strategy of being "Fustest with the Mostest" is not a business case but Humboldt's University of Berlin. Humboldt was actually not a bit interested in a university, as such. It was for him the means to create a new and different *political* order, which would be neither the absolute monarchy of the eighteenth century nor the democracy of the French Revolution in which the bourgeoisie ruled. Rather, it would be a balanced system, in which a totally apolitical professional civil service and an equally apolitical professional officer corps, recruited and promoted strictly by merit, would be autonomous in their very narrow spheres. These people—today we would call them technocrats—would have limited tasks and would be under the strict supervision of an independent professional judiciary. But within these limits they would be the masters. There would then be two spheres of individual freedom for the bourgeoisie, a moral and cultural one, and an economic one.

Humboldt had presented this concept earlier in book form.* After the total defeat of the Prussian monarchy by Napoleon in 1806, the collapse paralyzed all the forces that would otherwise have stopped Humboldt—the king, the aristocracy, the military. He ran with the opportunity and founded the University of Berlin as the main carrier of his political concepts, with brilliant success. The University of Berlin did indeed create the peculiar political structure the Germans in the nineteenth century called the *"Rechtsstaat"* (the Lawful State), in which an autonomous and self-governing elite of civil servants and general staff officers was in full control of the political and military sphere; an autonomous and self-governing elite of educated people *("die Gebildeten Staende")* organized around self-governing universities provided a "liberal" cultural sphere; and in which there was an autonomous and largely unrestricted economy. This structure first gave Prussia the moral and cultural, and soon thereafter the political and economic ascendancy in Germany. Both leadership in Europe and admiration outside of it followed in short order, especially on the part of the British and the Americans for whom the Germans, until 1890 or so, were the cultural and intellectual models. All this was exactly what Humboldt in the hour of darkest defeat and total despair had envisaged and aimed at. Indeed, he spelled out his aims clearly in the prospectus and the charter of his university.

Perhaps because "Fustest with the Mostest" must aim at creating something truly new, something truly different, nonexperts and outsiders seem to do as well as the experts, in fact, often better. Hoffmann-LaRoche, for instance, did not owe its strategy to chemists, but to a musician who had married the granddaughter of the company's foun-

---

*Under the title *The Limits on the Effectiveness of Government (Die Grenzen der Wirksamkeit des Staates)*, one of the very few original books on political philosophy ever written by a German.

der and needed more money to support his orchestra than the company then provided through its meager dividends. To this day the company has never been managed by chemists, but always by financial men who have made their career in a major Swiss bank. Wilhelm von Humboldt himself was a diplomat with no earlier ties to academia or experience in it. The DuPont top management people were businessmen rather than chemists and researchers. And while the Brothers Mayo were well-trained surgeons, they were totally outside the medical establishment of the time and isolated from it.

Of course, there are also the true "insiders," Dr. Wang or the people at 3M or the young computer engineers who designed the Apple computer. But when it comes to being "Fustest with the Mostest," the outsider may have an advantage. He does not know what everybody within the field knows, and therefore does not know what cannot be done.

II

The strategy of being "Fustest with the Mostest" has to hit right on target or it misses altogether. Or, to vary the metaphor, being "Fustest with the Mostest" is very much like a moon shot: a deviation of a fraction of a minute of the arc and the missile disappears into outer space. And once launched, the "Fustest with the Mostest" strategy is difficult to adjust or to correct.

To use this strategy, in other words, requires thought and careful analysis. The entrepreneur of so much of the popular literature or of Hollywood movies, the person who suddenly has a "brilliant idea" and rushes off to put it into effect, is not going to succeed with it. In fact, for this strategy to succeed at all, the innovation must be based on a careful and deliberate attempt to exploit one of the major opportunities for innovation that were discussed in Chapters 3 to 9.

There is, for instance, no better example of exploiting a *change in perception* than Humboldt's University of Berlin. The French Revolution with its Terror, followed by Napoleon's ruthless wars of conquest, had left the educated bourgeoisie disillusioned with politics; and yet they also quite clearly would have rejected any attempt to move the clock back and return to the absolute monarchy of the eighteenth century, let alone to feudalism. They needed a "liberal" but apolitical sphere, coupled with an apolitical government based on the same principles of law and education in which they themselves believed. And all of them at the time were followers of Adam Smith, whose *Wealth of Nations* was probably the most widely read and most highly respected political book of the period. It was this which Humboldt's political structure exploited and which his plan for the University of Berlin translated into institutional reality.

Wang's word processor brilliantly exploited a process need. By the 1970s the fear of the computer that had been rampant in offices only a little while earlier was beginning to be replaced by the question, "And what will the computer do for *me?*" By that time, office workers had become familiar with the computer in such activities as making payroll or controlling inventories; they also by that time had acquired office copiers so that the paperload in every office was going up very sharply. Wang's word processor then addressed itself to the one remaining nonautomated chore, a chore every office worker hated: rewriting letters, speeches, reports, manuscripts to embody minor changes, and having to do so again and again.

Hoffmann-LaRoche, in picking the vitamins in the early twenties, exploited new knowledge. The musician who laid down its strategy understood the "structure of scientific revolutions" a full thirty years before a philosopher, Thomas Kuhn, wrote the celebrated book by that title. He understood that a new basic theorem in science, even though buttressed by enough evidence to make it impossible to reject, will still not be accepted by a majority of scientists should it conflict with basic theorems they have grown up with and hold as articles of faith. They pay no attention to it for a long time, until the old "paradigm," the old basic theory, becomes totally untenable. And during that time those who accept the new theorem and run with it have the field all to themselves.

Only with such a base in careful analysis can the strategy of being "Fustest with the Mostest" possibly succeed.

Even then, it requires extreme concentration of effort. There has to be one clear-cut goal and all efforts have to be focused on it. And when this effort begins to produce results, the innovator has to be ready to mobilize resources massively. As soon as DuPont had a usable synthetic fiber—long before the market had begun to respond to it—the company built large factories and bombarded both textile manufacturers and the general public with advertisements, trial presentations, and samples.

Then, after the innovation has become a successful business, the work really begins. Then the strategy of "Fustest with the Mostest" demands substantial and continuing efforts to retain a leadership position; otherwise, all one has done is create a market for a competitor. The innovator has to run even harder now that he has leadership than he ran before and to continue his innovative efforts on a very large scale. The research budget must be higher *after* the innovation has successfully been accomplished than it was before. New uses have to be found; new customers must be identified, and persuaded to try the new materials. Above all, the entrepreneur who has succeeded in being "Fustest with the Mostest" has to make his product or his process obsolete before a competitor can do it. Work on the successor to the successful product or process has to start immediately, with the same concentration of effort and the same investment of resources that led to the initial success.

Finally, the entrepreneur who has attained leadership by being "Fustest with the Mostest" has to be the one who systematically cuts the price of his own product or process. To keep prices high simply holds an umbrella over potential competitors and encourages them (on this, see the next chapter, "Hit Them Where They Ain't").

This was established by the longest-lived private monopoly in economic history, the Dynamite Cartel, founded by Alfred Nobel after his invention of dynamite. The Dynamite Cartel maintained a worldwide monopoly until World War I and even beyond, long after the Nobel patents had expired. It did this by cutting price every time demand rose by 10 to 20 percent. By that time, the companies in the cartel had fully depreciated the investment they had had to make to get the additional production. This made it unattractive for any potential competitor to build new dynamite factories, while the cartel itself maintained its profitability. It is no accident that DuPont has consistently followed this policy in the United States, for the DuPont Company was the American member of the Dynamite Cartel. But Wang has done the same with respect to the word processor, Apple with respect to its computers, and 3M with respect to all its products.

III

These are all success stories. They do *not* show how risky the strategy of being "Fustest with the Mostest" actually is. The failures disappeared. Yet we know that for everyone who succeeds with this strategy, many more fail. There is only one chance with the "Fustest with the Mostest" strategy. If it does not work right away, it is a total failure.

Everyone knows the old Swiss story of Wilhelm Tell the archer, whom the tyrant promised to pardon if he succeeded in shooting an apple off his son's head on the first try. If he failed, he would either kill the child or be killed himself. This is exactly the situation of the entrepreneur in the "Fustest with the Mostest" strategy. There can be no "almost-success" or "near-miss." There is only success or failure.

Even the successes may be perceived only by hindsight. At least we know that in two of the examples failure was very close; a combination of luck and chance saved them.

Nylon only succeeded because of a fluke. There was no market for a synthetic fiber in the mid-thirties. It was far too expensive to compete with cotton and rayon, the cheap fibers of the time, and was actually even more expensive than silk, the luxury fiber which the Japanese in the severe depression of the late thirties had to sell for whatever price they could get. What saved Nylon was the outbreak of World War II, which stopped Japanese silk exports. By the time the Japanese could start up their silk industry again, around 1950 or so, Nylon was firmly entrenched, with its cost and price down to a fraction of what both had been in the late thirties. The story of 3M's best known product, Scotch Tape, was told earlier. Again, but for pure accident, Scotch Tape would have been a failure.

The strategy of being "Fustest with the Mostest" is indeed so risky that an entire major strategy—the one that will be discussed in the next chapter under the heading Creative Imitation—is based on the assumption that being "Fustest with the Mostest" will fail far more often than it can possibly succeed. It will fail because the will is lacking. It will fail because efforts are inadequate. It will fail because, despite successful innovation, not enough resources are deployed, are available, or are being put to work to exploit success, and so on. While the strategy is indeed highly rewarding when successful, it is much too risky and much too difficult to be used for anything but major innovations, for creating a new political order as Humboldt successfully did, or a whole new field of therapy as Hoffmann-LaRoche did with the vitamins, or a new approach to medical diagnosis and practice as the Mayo Brothers set out to do. In effect, it fits a fairly small minority of innovations. It requires profound analysis and a genuine understanding of the sources of innovation and of their dynamics. It requires an extreme concentration of effort and substantial resources. In most cases alternative strategies are available and preferable—not primarily because they carry less risk, but because for most innovations the opportunity is not great enough to justify the cost, the effort, and the investment of resources required for the "Fustest with the Mostest" strategy.

17

"Hit Them Where They Ain't"

Two completely different entrepreneurial strategies were summed up by another battle-winning Confederate general in America's Civil War, who said: "Hit Them Where They Ain't." They might be called creative imitation and entrepreneurial judo, respectively.

I

CREATIVE IMITATION

Creative imitation* is clearly a contradiction in terms. What is creative must surely be original. And if there is one thing imitation is not, it is "original." Yet the term fits. It describes a strategy that is "imitation" in its substance. What the entrepreneur does is something somebody else has already done. But it is "creative" because the entrepreneur applying the strategy of "creative imitation" understands what the innovation represents better than the people who made it and who innovated.

The foremost practitioner of this strategy and the most brilliant one is IBM. But it is also very largely the strategy that Procter & Gamble has been using to obtain and maintain leadership in the soap, detergent, and toiletries markets. And the Japanese Hattori Company, whose Seiko watches have become the world's leader, also owes its domination of the market to creative imitation.

In the early thirties IBM built a high-speed calculating machine to do calculations for the astronomers at New York's Columbia University. A few years later it built a machine that was already designed as a computer—again, to do astronomical calculations, this time at Harvard. And by the end of World War II, IBM had built a real computer

*The term was coined by Theodore Levitt of the Harvard Business School.

—the first one, by the way, that had the features of the true computer: a "memory" and the capacity to be "programmed." And yet there are good reasons why the history books pay scant attention to IBM as a computer innovator. For as soon as it had finished its advanced 1945 computer—the first computer to be shown to a lay public in its showroom in midtown New York, where it drew immense crowds—IBM abandoned its own design and switched to the design of its rival, the ENIAC developed at the University of Pennsylvania. The ENIAC was far better suited to business applications such as payroll, only its designers did not see this. IBM structured the ENIAC so that it could be manufactured and serviced and could do mundane "numbers crunching." When IBM's version of the ENIAC came out in 1953, it at once set the standard for commercial, multipurpose, mainframe computers.

This is the strategy of "creative imitation." It waits until somebody else has established the new, but only "approximately." Then it goes to work. And within a short time it comes out with what the new really should be to satisfy the customer, to do the work customers want and pay for. The creative imitation has then set the standard and takes over the market.

IBM practiced creative imitation again with the personal computer. The idea was Apple's. As described earlier (in Chapter 3), everybody at IBM "knew" that a small, freestanding computer was a mistake—uneconomical, far from optimal, and expensive. And yet it succeeded. IBM immediately went to work to design a machine that would become the standard in the personal computer field and dominate or at least lead the entire field. The result was the PC. Within two years it had taken over from Apple leadership in the personal computer field, becoming the fastest-selling brand and the standard in the field.

Procter & Gamble acts very much the same way in the market for detergents, soaps, toiletries, and processed foods.

When semiconductors became available, everyone in the watch industry knew that they could be used to power a watch much more accurately, much more reliably, and much more cheaply than traditional watch movements. The Swiss soon brought out a quartz-powered digital watch. But they had so much investment in traditional watchmaking that they decided on a gradual introduction of quartz-powered digital watches over a long period of time, during which these new timepieces would remain expensive luxuries.

Meanwhile, the Hattori Company in Japan had long been making conventional watches for the Japanese market. It saw the opportunity and went in for creative imitation, developing the quartz-powered digital watch as the standard timepiece. By the time the Swiss had woken up, it was too late. Seiko watches had become the world's best-sellers, with the Swiss almost pushed out of the market.

Like being "Fustest with the Mostest," creative imitation is a strategy aimed at market or industry leadership, if not at market or industry dominance. But it is much less risky. By the time the creative imitator moves, the market has been established and the new venture has been accepted. Indeed, there is usually more demand for it than the original innovator can easily supply. The market segmentations are known or at least knowable. By then, too, market research can find out what customers buy, how they buy, what constitutes value for them, and so on. Most of the uncertainties that abound when the first innovator appears have been dispelled or can at least be analyzed and studied. No one has to explain any more what a personal computer or a digital watch are and what they can do.

Of course, the original innovator may do it right the first time, thus closing the door to creative imitation. There is the risk of an innovator bringing out and doing the right job with vitamins as Hoffmann-LaRoche did, or with Nylon as did DuPont, or as Wang did with the word processor. But the number of entrepreneurs engaging in creative imitation, and their substantial success, indicates that perhaps the risk of the first innovator's preempting the market by getting it right is not an overwhelming one.

Another good example of creative imitation is Tylenol, the "non-aspirin aspirin." This case shows more clearly than any other I know what the strategy consists of, what its requirements are, and how it works.

Acetaminophen (the substance that is sold under the Tylenol brand name in the U.S.) had been used for many years as a painkiller, but until recently it was available in the United States only by prescription. Until recently also, aspirin, the much older pain-killing substance, was considered perfectly safe and had the pain-relief market to itself. Acetaminophen is a less potent drug than aspirin. It is effective as a painkiller but has no anti-inflammatory effect and also no effect on blood coagulation. Because of this it is free from the side effects, especially gastric upsets and stomach bleeding, which aspirin can cause, particularly if used in large doses and over long periods of time for an illness like arthritis.

When acetaminophen became available without prescription, the first brand on the market was presented and promoted as a drug for those who suffered side effects from aspirin. It was eminently successful, indeed, far more successful than its makers had anticipated. But it was this very success that created the opportunity for creative imitation. Johnson & Johnson realized that there was a market for a drug that *replaced* aspirin as the painkiller of choice, with aspirin confined to the fairly small market where anti-inflammatory and blood coagulation effects were needed. From the start Tylenol was promoted as the safe, *universal* painkiller. Within a year or two it had the market.

Creative imitation, these cases show, does not exploit the failure of the pioneers as failure is commonly understood. On the contrary, the pioneer must be successful. The Apple computer was a great success story, and so was the acetaminophen brand that Tylenol ultimately pushed out of market leadership. But the original innovators failed to understand their success. The makers of the Apple were product-focused rather than user-focused, and therefore offered additional hardware where the user needed programs and software. In the Tylenol case, the original innovators failed to realize what their own success meant.

The creative innovator exploits the success of others. Creative imitation is not "innovation" in the sense in which the term is most commonly understood. The creative imitator does not invent a product or service; he perfects and positions it. In the form in which it has been introduced, it lacks something. It may be additional product features. It may be segmentation of product or services so that slightly different versions fit slightly different markets. It might be proper positioning of the product in the market. Or creative imitation supplies something that is still lacking.

The creative imitator looks at products or services from the viewpoint of the customer. IBM's personal computer is practically indistinguishable from the Apple in its technical features, but IBM from the beginning offered the customer programs and software. Apple maintained traditional computer distribution through specialty stores. IBM —in a radical break with its own traditions—developed all kinds of distribution channels, specialty stores, major retailers like Sears, Roebuck, its own retail stores, and so on. It made it easy for the consumer to buy and it made it easy for the consumer to use the product. These, rather than hardware features, were the "innovations" that gave IBM the personal computer market.

All told, creative imitation starts out with markets rather than with products, and with customers rather than with producers. It is both market-focused and market-driven.

These cases show what the strategy of creative imitation requires:

It requires a rapidly growing market. Creative imitators do not succeed by taking away customers from the pioneers who have first introduced a new product or service; they serve markets the pioneers have created but do not adequately service. Creative imitation satisfies a demand that already exists rather than creating one.

The strategy has its own risks, and they are considerable. Creative imitators are easily tempted to splinter their efforts in the attempt to hedge their bets. Another danger is to misread the trend and imitate creatively what then turns out not to be the winning development in the marketplace.

IBM, the world's foremost creative imitator, exemplifies these dangers. It has successfully imitated every major development in the office-automation field. As a result, it has the leading product in every single area. But because they originated in imitation, the products are so diverse and so little compatible with one another that it is all but impossible to build an integrated, automated office out of IBM building blocks. It is thus still doubtful that IBM can maintain leadership in the automated office and provide the integrated system for it. Yet this is where the main market of the future is going to be in all probability. And this risk, *the risk of being too clever,* is inherent in the creative imitation strategy.

Creative imitation is likely to work most effectively in high-tech areas for one simple reason: high-tech innovators are least likely to be market-focused, and most likely to be technology- and product-focused. They therefore tend to misunderstand their own success and to fail to exploit and supply the demand they have created. But as acetaminophen and the Seiko watch show, they are by no means the only ones to do so.

Because creative imitation aims at market dominance, it is best suited to a major product, process, or service: the personal computer, the worldwide watch market, or a market as large as that for pain relief. But the strategy requires less of a market than being "Fustest with the Mostest." It carries less risk. By the time creative imitators go to work, the market has already been identified and the demand has already been created. What it lacks in risk, however, creative imitation makes up for in its requirements for alertness, for flexibility, for willingness to accept the verdict of the market, and above all, for hard work and massive efforts.

II

ENTREPRENEURIAL JUDO

In 1947, Bell Laboratories invented the transistor. It was at once realized that the transistor was going to replace the vacuum tube, especially in consumer electronics such as the radio and the brand-new television set. Everybody knew this; but nobody did anything about it. The leading manufacturers—at that time they were all Americans— began to study the transistor and to make plans for conversion to the transistor "sometime around 1970." Till then, they proclaimed, the transistor "would not be ready." Sony was practically unknown outside of Japan and was not even in consumer electronics at the time. But Akio Morita, Sony's president, read about the transistor in the newspapers. As a result, he went to the United States and bought a license for the new transistor from Bell Labs for a ridiculous sum, all of $25,000. Two years later, Sony brought out the first portable transistor radio, which weighed less than one-fifth of comparable vacuum tube radios on the market, and cost less than one-third. Three years later, Sony had the market for cheap radios in the United States; and five years later, the Japanese had captured the radio market all over the world.

Of course, this is a classic case of the rejection of the unexpected success. The Americans rejected the transistor because it was "not invented here," that is, not invented by the electrical and electronic leaders, RCA and G.E. It is a typical example of pride in doing things the hard way. The Americans were so proud of the wonderful radios of those days, the great Super Heterodyne sets that were such marvels of craftsmanship. Compared to them, they thought silicon chips low grade, if not indeed beneath their dignity.

But Sony's success is not the real story. How do we explain that the Japanese repeated this same strategy again and again, and always with success, always surprising the Americans? They repeated it with television sets and digital watches and hand-held calculators. They repeated it with copiers when they moved in and took away a large share of the market from the original inventor, the Xerox Company. The Japanese, in other words, have been successful again and again in practicing "entrepreneurial judo" against the Americans.

But so did MCI and Sprint when they used the Bell Telephone System's (AT&T) own pricing to take away from the Bell System a very large part of the long-distance business (see Chapter 6). So did ROLM when it used Bell System's policies against it to take away a large part of the private branch exchange (PBX) market. And so did Citibank when it started a consumer bank in Germany, the *"Familienbank"* (Family Bank), which within a few short years came to dominate German consumer finance.

The German banks knew that ordinary consumers had obtained purchasing power and had become desirable clients. They went through the motions of offering consumers banking services. But they really did not want them. Consumers, they felt, were beneath the dignity of a major bank, with its business customers and its rich investment clients. If consumers needed an account at all, they should have it with the postal savings bank. Whatever their advertisements said to the contrary, the banks made it abundantly clear when consumers came into the august offices of the local branch that they had little use for them.

This was the opening Citibank exploited when it founded its German *Familienbank,* which catered to none but individual consumers, designed the services consumers needed, and made it easy for consumers to do business with a bank. Despite the tremendous strength of the German banks and their pervasive presence in a country where there is a branch of a major bank on the corner of every downtown street, Citibank's *Familienbank* attained dominance in the German consumer banking business within five years or so.

All these newcomers—the Japanese, MCI, ROLM, Citibank—practiced "entrepreneurial judo." Of the entrepreneurial strategies, especially the strategies aimed at obtaining leadership and dominance in an industry or a market, entrepreneurial judo is by all odds the least risky and the most likely to succeed.

Every policeman knows that a habitual criminal will always commit his crime the same way—whether it is cracking a safe or entering a building he wants to loot. He leaves behind a "signature," which is as individual and as distinct as a fingerprint. And he will not change that signature even though it leads to his being caught time and again.

But it is not only the criminal who is set in his habits. All of us are. And so are businesses and industries. The habit will be persisted in even though it leads again and again to loss of leadership and loss of market. The American manufacturers persisted in the habits that enabled the Japanese to take over their market again and again.

If the criminal is caught, he rarely accepts that his habit has betrayed him. On the contrary, he will find all kinds of excuses—and continue the habit that led to his being captured. Similarly, businesses that are being betrayed by their habits will not admit it and will find all kinds of excuses. The American electronics manufacturers, for instance, attribute the Japanese successes to "low labor costs" in Japan. Yet the few American manufacturers that have faced up to reality, for example, RCA and Magnavox in television sets, are able to turn out in the United States products at prices competitive with those of the Japanese, and competitive also in quality, despite their paying American wages and union benefits. The German banks uniformly explain the success of Citibank's *Familienbank* by its taking risks they themselves would not touch. But *Familienbank* has lower credit losses with consumer loans than the German banks, and its lending requirements are as strict as those of the Germans. The German banks know this, of course. Yet they keep on explaining away their failure and *Familienbank*'s success. This is typical. And it explains why the same strategy—the same entrepreneurial judo—can be used over and over again.

There are in particular five fairly common bad habits that enable newcomers to use entrepreneurial judo and to catapult themselves into a leadership position in an industry against the entrenched, established companies.

1. The first is what American slang calls "NIH" ("Not Invented Here"), the arrogance that leads a company or an industry to believe that something new cannot be any good unless they themselves thought of it. And so the new invention is spurned, as was the transistor by the American electronics manufacturers.

2. The second is the tendency to "cream" a market, that is, to get the high-profit part of it.

This is basically what Xerox did and what made it an easy target for the Japanese imitators of its copying machines. Xerox focused its strategy on the big users, the buyers of large numbers of machines or of expensive, high-performance machines. It did not reject the others; but it did not go after them. In particular, it did not see fit to give them service. In the end it was dissatisfaction with the service—or rather, with the lack of service—Xerox provided for its smaller customers that made them receptive to competitors' machines.

"Creaming" is a violation of elementary managerial and economic precepts. It is always punished by loss of market.

Xerox was resting on its laurels. They were indeed substantial and well earned, but no business ever gets paid for what it did in the past. "Creaming" attempts to get paid for past contributions. Once a business gets into that habit, it is likely to continue in it and thus continue to be vulnerable to entrepreneurial judo.

3. Even more debilitating is the third bad habit: the belief in "quality." "Quality" in a product or service is not what the supplier puts in. It is what the customer gets out and is willing to pay for. A product is not "quality" because it is hard to make and costs a lot of money, as manufacturers typically believe. That is incompetence. Customers pay only for what is of use to them and gives them value. Nothing else constitutes "quality."

The American electronics manufacturers in the 1950s believed that their products with all those wonderful vacuum tubes were "quality" because they had put in thirty years of effort making radio sets more complicated, bigger, and more expensive. They considered the product to be "quality" because it needed a great deal of skill to turn out, whereas a transistor radio is simple and can be made by unskilled labor on the assembly line. But in consumer terms, the transistor radio is clearly far superior "quality." It weighs much less so that it can be taken on a trip to the beach or to a picnic. It rarely goes wrong; there are no tubes to replace. It costs a great deal less. And in range and fidelity it very soon surpassed even the most magnificent Super Heterodyne with sixteen vacuum tubes, one of which always burned out just when needed.

4. Closely related to both "creaming" and "quality" is the fourth bad habit, the delusion of the "premium" price. A "premium" price is always an invitation to the competitor.

For two hundred years, since the time of J. B. Say in France and of David Ricardo in England in the early years of the nineteenth century, economists have known that the only way to get a higher profit margin, except through a monopoly, is through lower costs. The attempt to achieve a higher profit margin through a higher price is always self-defeating. It holds an umbrella over the competitor. What looks like higher profits for the established leader is in effect a subsidy to the newcomer who, in a very few years, will unseat the leader and claim the throne for himself. "Premium" prices, instead of being an occasion for joy—and a reason for a higher stock price or a higher price/earnings multiple—should always be considered a threat and dangerous vulnerability.

Yet the delusion of higher profits to be achieved through "premium" prices is almost universal, even though it always opens the door to entrepreneurial judo.

5. Finally, there is a fifth bad habit that is typical of established businesses and leads to their downfall—Xerox is a good example. They maximize rather than optimize. As the market grows and develops, they try to satisfy every single user through the same product or service.

A new analytical instrument to test chemical reaction is being introduced, for instance. At first its market is quite limited, let's say to industrial laboratories. But then university laboratories, research institutes, and hospitals all begin to buy the instrument, but each wants something slightly different. And so the manufacturer puts in one feature to satisfy this customer, then another one to satisfy that customer, and so on, until what started out as a simple instrument has become complicated. The manufacturer has maximized what the instrument can do. As a result, the instrument no longer satisfies anyone. For, by trying to satisfy everybody, one always ends up satisfying nobody. The instrument also has become expensive, as well as being hard to use and hard to maintain. But the manufacturer is proud of the instrument; indeed, his full-page advertisement lists sixty-four different things it can do.

This manufacturer will almost certainly become the victim of entrepreneurial judo. What he thinks is his very strength will be turned against him. The newcomer will come in with an instrument designed to satisfy one of the markets, the hospital, for instance. It will not contain a single feature the hospital people do not need, and do not need every day. But everything the hospital needs will be there and with higher performance capacity than the multipurpose instrument can possibly offer. The same manufacturer will then bring out a model for the research laboratory, for the government laboratory, for industry —and in no time at all the newcomer will have taken away the markets with instruments that are specifically designed for their users, instruments that optimize rather than maximize.

Similarly, when the Japanese came in with their copiers in competition with Xerox, they designed machines that fitted specific groups of users—for example, the small office, whether that of the dentist, the doctor, or the school principal. They did not try to match the features of which the Xerox people themselves were the proudest, such as the speed of the machine or the clarity of the copy. They gave the small office what the small office needed most, a simple machine at a low cost. And once they had established themselves in that market, they then moved in on the other markets, each with a product designed to serve optimally a specific market segment.

Sony similarly first moved into the low end of the radio market, the market for cheap portables with limited range. Once it had established itself there, it moved in on the other market segments.

Entrepreneurial judo aims first at securing a beachhead, and one which the established leaders either do not defend at all or defend only halfheartedly—the way the Germans did not counterattack when Citibank established its *Familienbank*. Once that beachhead has been secured, that is, once the newcomers have an adequate market and an adequate revenue stream, they then move on to the rest of the "beach" and finally to the whole "island." In each case, they repeat the strategy. They design a product or a service which is specific to a given market segment and optimal for it. And the established leaders hardly ever beat them to this game. Hardly ever do the established leaders manage to change their own behavior before the newcomers have taken over the leadership and acquired dominance.

There are three situations in which the entrepreneurial judo strategy is likely to be particularly successful.

The first is the common situation in which the established leaders refuse to act on the unexpected, whether success or failure, and either overlook it altogether or try to brush it aside. This is what Sony exploited.

The second situation is the Xerox situation. A new technology emerges and grows fast. But the innovators who have brought to the market the new technology (or the new service) behave like the classical "monopolists": they use their leadership position to "cream" the market and to get "premium" prices. They either do not know or refuse to acknowledge what has been amply proven: that a leadership position, let alone any kind of monopoly, can only be maintained if the leader behaves as a "benevolent monopolist" (the term is Joseph Schumpeter's).

A benevolent monopolist cuts his prices before a competitor can cut them. And he makes his product obsolete and introduces new product before a competitor can do so. There are enough examples of this around to prove the validity of the thesis. It is the way in which the DuPont Company has acted for many years and in which the American Bell Telephone System (AT&T) used to act before it was overcome by the inflationary problems of the 1970s. But if the leader uses his leadership position to raise prices or to raise profit margins except by lowering his cost, he sets himself up to be knocked down by anyone who uses entrepreneurial judo against him.

Similarly, the leader in a rapidly growing new market or new technology who tries to maximize rather than to optimize will soon make himself vulnerable to entrepreneurial judo.

Finally, entrepreneurial judo works as a strategy when market or industry structure changes fast—which is the *Familienbank* story. As Germany became prosperous in the fifties and sixties, ordinary people became customers for financial services beyond the traditional savings account or the traditional mortgage. But the German banks stuck to their old markets.

Entrepreneurial judo is always market-focused and market-driven. The starting point may be technology, as it was when Akio Morita traveled to the United States from a Japan that had barely emerged from the destruction of World War II to acquire a transistor license. Morita looked at the market segment which the existing technology satisfied the least, simply because of the weight and fragility of vacuum tubes: the market for portables. He then designed the right radio for that market, a market of young people with little money but also fairly simple demands with respect to range of the instrument and to quality of sound, a market, in other words, that the old technology simply could not adequately serve.

Similarly, the long-distance discounters in the United States who saw the opportunity to buy from the Bell Telephone System wholesale and to resell retail, designed a service first for the fairly modest number of substantial businesses that were too small to build their own longdistance system but large enough to have heavy long-distance bills. Only after they had secured a substantial share of that market did they move out and try to go after both the very big and the small users.

To use the entrepreneurial judo strategy, one starts out with an analysis of the industry, the producers and the suppliers, their habits, especially their bad habits, and their policies. But then one looks at the markets and tries to pinpoint the place where an alternative strategy would meet with the greatest success and the least resistance.

Entrepreneurial judo requires some degree of genuine innovation. It is, as a rule, not good enough to offer the same product or the same service at lower cost. There has to be something that distinguishes it from what already exists. When the ROLM Company offered a private branch exchange—a switchboard for business and office users—in competition with AT&T, it built in additional features designed around a small computer. These were not high-tech, let alone new inventions. Indeed, AT&T itself had designed similar features. But AT&T did not push them—and ROLM did. Similarly, when Citibank went into Germany with the *Familienbank*, it put in some innovative services which German banks as a rule did not offer to small depositors, such as travelers checks or tax advice.

It is not enough, in other words, for the newcomer simply to do as good a job as the established leader at a lower cost or with better service. The newcomers have to make themselves distinct.

Like being "Fustest with the Mostest" and creative imitation, entrepreneurial judo aims at obtaining leadership position and eventually dominance. But it does not do so by competing with the leaders—or at least not where the leaders are aware of competitive challenge or worried about it. Entrepreneurial judo "Hits Them Where They Ain't."

18

Ecological Niches

The entrepreneurial strategies discussed so far, being "Fustest with the Mostest," creative imitation, and entrepreneurial judo, all aim at market or industry leadership, if not at dominance. The "ecological niche" strategy aims at control. The strategies discussed earlier aim at positioning an enterprise in a large market or a major industry. The ecological niche strategy aims at obtaining a practical monopoly in a small area. The first three strategies are competitive strategies. The ecological niche strategy aims at making its successful practitioners immune to competition and unlikely to be challenged. Successful practitioners of "Fustest with the Mostest," creative imitation, and entrepreneurial judo become big companies, highly visible if not household words. Successful practitioners of the ecological niche take the cash and let the credit go. They wallow in their anonymity. Indeed, in the most successful of the ecological niche strategies, the whole point is to be so inconspicuous, despite the product's being essential to a process, that no one is likely to try to compete.

There are three distinct niche strategies, each with its own requirements, its own limitations, and its own risks:

- the toll-gate strategy;
- the specialty skill strategy; and
- the specialty market strategy.

I

THE TOLL-GATE STRATEGY

Earlier, in Chapter 4, I discussed the strategy of the Alcon Company, which developed an enzyme to eliminate the one feature of the standard surgical operation for senile cataracts that went counter to the rhythm and the logic of the process. Once this enzyme had been developed and patented, it had a "toll-gate" position. No eye surgeon would do without it. No matter what Alcon charged for the teaspoonful of enzyme that was needed for each cataract operation, the cost was insignificant in relation to the total cost of the operation. I doubt that any eye surgeon or any hospital ever even inquired what the stuff cost. The total market for this particular preparation was so small—maybe $50 million dollars a year worldwide—that it clearly would not have been worth anybody's while to try to develop a competing product. There would not have been one additional cataract operation in the world just because this particular enzyme had become cheaper. All that potential competitors could possibly do, therefore, would have been to knock down the price for everybody, without deriving much benefit for themselves.

A very similar toll-gate position has been occupied for many years by a medium-sized company which, fifty or sixty years ago, developed a blowout protector for oil wells. The cost of drilling an oil well may run into many millions. One blowout will destroy the entire well and everything that has been invested in it. The blowout protector, which safeguards the well while being drilled, is thus cheap insurance, no matter what its price. Again, the total market is so limited as to make it unattractive for any would-be competitor. Lowering the price of blowout protectors, which constitute maybe 1 percent of the total cost of a deep well, could not possibly stimulate anyone to drill more wells. Competition could only degrade the price without increasing the demand.

Another example of a toll-gate strategy is Dewey & Almy—now a division of W. R. Grace. This company developed a compound to seal tin cans in the 1930s. The seal is an essential ingredient of the can: if a can goes bad, it can cause catastrophic damage. One death from one case of botulism in a can can easily destroy a food packer. A can-sealing compound that offers protection against spoilage is therefore cheap at any price. And yet the cost of sealing—a fraction of a cent at best—is so insignificant to both the cost of the total can and the risk of spoilage that nobody is much concerned about it. What matters is performance, not cost. Again, the total market, while larger than that for enzymes in cataract operations or for blowout protectors, is still a limited one. And lowering the price for can-sealing compound is quite unlikely to increase the demand by a single can.

The toll-gate position is thus in many ways the most desirable position a company can occupy. But it has stringent requirements. The product has to be essential to a process. The risk of not using it—the risk of losing an eye, losing an oil well, or spoilage in a tin can—must be infinitely greater than the cost of the product. The market must be so limited that whoever occupies it first preempts it. It must be a true "ecological niche" which one species fills completely, and which at the same time is small and discreet enough not to attract rivals.

Such toll-gate positions are not easily found. Normally they occur only in an incongruity situation (cf. Chapter 4). The incongruity, as in the case of Alcon's enzyme, might be an incongruity in the rhythm or the logic of a process. Or, as in the case of the blowout protector or the can-sealing compound, it might be an incongruity between economic realities—between the cost of malfunction and the cost of adequate protection.

The toll-gate position also has severe limitations and serious risks. It is basically a static position. Once the ecological niche has been occupied, there is unlikely to be much growth. There is nothing the company that occupies the toll-gate position can do to increase its business or to control it. No matter how good its product or how cheap, the demand is dependent upon the demand for the process or product to which the toll-gate product furnishes an ingredient.

This may not be too important for Alcon. Cataracts can be assumed to be impervious to economic fluctuations, whether boom or depression. But the company making blowout protectors had to invest enormous amounts of money in new plants when oil drilling skyrocketed in 1973, and again after the 1979 petroleum panic. It suspected that the boom could not last; yet it had to make the investments even though it was reasonably sure it could never earn them back. Not to have done so would have meant losing its market irretrievably. Equally, it was powerless when, a few years later, the oil boom collapsed and oil drilling shrank by 80 percent within twelve months, and with it orders for oil-drilling equipment.

Once the toll-gate strategy has attained its objective, the company is "mature." It can only grow as fast as its end users grow. But it can go down fast. It can become obsolete almost overnight if someone finds a different way of satisfying the same end use. Dewey & Almy, for instance, has no defense against the replacement of tin cans by other container materials such as glass, paper, or plastics, or by other methods of preserving food such as freezing and irradiation.

And the toll-gate strategist must never exploit his monopoly. He must not become what the Germans call a *Raubritter* (the English "robber baron" does not mean quite the same thing) who robbed and raped the hapless travelers as they passed through the mountain defiles and river gorges atop of which perched his castle. He must not abuse his monopoly to exploit, to extort, to maltreat his customers. If he does, the users will put another supplier into business, or they will switch to less effective substitutes which they can then control.

The right strategy is the one Dewey & Almy has successfully pursued for more than forty years now. It offers its users, especially those in the Third World, extensive technical service, teaches their people, and designs new and better canning and can-sealing machinery for them to use with the Dewey & Almy sealing compounds. Yet it also constantly upgrades the compounds.

The toll-gate position might be impregnable—or nearly so. But it can only control within a narrow radius. Alcon tried to overcome this limitation by diversifying into all kinds of consumer products for the eye: artificial tears, contact lens fluids, anti-allergic eyedrops, and so on. This was successful insofar as it made the company attractive to one of the leading consumer goods multinationals, the Swiss Nestlé Company, which bought out Alcon for a very substantial sum. To the best of my knowledge, Alcon is the only toll-gate company of this kind that succeeded in establishing itself in markets outside its original position and with products that were different in their economic characteristics. But whether this diversification into highly competitive consumer markets of which the company knew very little was profitable, is not known.

II

THE SPECIALTY SKILL

Everybody knows the major automobile nameplates. But few people know the names of the companies that supply the electrical and lighting systems for these cars, and yet there are far fewer such systems than there are automobile nameplates: in the United States, the Delco group of GM; in Germany, Robert Bosch; in Great Britain, Lucas; and so on. Practically no one outside of the automobile industry knows that one firm, A. O. Smith of Milwaukee, has for decades been making every single frame used in an American passenger car, nor that for decades another firm, Bendix, has made every single set of automotive brakes used by the American automobile industry.

By now these are all old and well-established firms, of course, but only because the automobile is itself an old industry. These companies established their controlling position when the industry was in its infancy, well before World War I. Robert Bosch, for instance, was a contemporary and friend of the two German auto pioneers, Carl Benz and Gottfried Daimler, and started his firm in the 1880s.

But once these companies had attained their controlling position in their specialty skill niche, they retained it. Unlike the toll-gate companies, theirs is a fairly large niche, yet it is still unique. It was obtained by developing high skill at a very early time. A. O. Smith developed what today would be called "automation" in making automobile frames during and shortly after World War I. The electrical system which Bosch in Germany designed for Mercedes staff cars around 1911 was so far advanced that it was put into general use even in luxury automobiles only after World War II. Delco in Dayton, Ohio, developed the self-starter before becoming a part of General Motors, that is, before 1914. Such specialized skills put these companies so far ahead in their field that it was hardly worth anybody's while to try to challenge them. They had become the "standard."

Specialty skill niches are by no means confined to manufacturing. Within the last ten years a few private trading firms, most of them in Vienna, Austria, have built a similar niche in what used to be called "barter" and is now called "counter-trade": taking goods from a developing importing country, Bulgarian tobacco or Brazilian-made irrigation pumps, in payment for locomotives, machinery, or pharmaceuticals exported by a company in a developed country. And much earlier, an enterprising German attained such a hold on one specialty skill niche that guidebooks for tourists are still called by his name, "Baedeker."

As these cases show, timing is of the essence in establishing a specialty skill niche. It has to be done at the very beginning of a new industry, a new custom, a new market, a new trend. Karl Baedeker published his first guidebook in 1828, as soon as the first steamships on the Rhine opened tourist travel to the middle classes. He then had the field virtually to himself until World War I made German books unacceptable in Western countries. The counter-traders of Vienna started around 1960, when such trade was still the rare exception, largely confined to the smaller countries of the Soviet Bloc (which explains why they are concentrated in Austria). Ten years later, when hard currencies had become scarce all through the Third World, they had honed their skills and become the "specialists."

To attain a specialty niche always requires something new, something added, something that is genuine innovation. There were guidebooks for travelers before Baedeker, but they confined themselves to the cultural scene—churches, sights, and so on. For practical details— the hotels, the tariff of the horse-drawn cabs, the distances, and the proper amount to tip—the traveling English milord relied on a professional, the courier. But the middle class had no courier, and that was Baedeker's opportunity. Once he had learned what information the traveler needed, how to get at it and to present it (the format he established is still the one many guidebooks follow), it would not have paid anyone to duplicate Baedeker's investment and build a competing organization.

In the early stages of a major new development, the specialty skill niche offers an exceptional opportunity. Examples abound. For many, many years there were only two companies in the United States making airplane propellers, for instance. Both had been started before World War I.

A specialty skill niche is rarely found by accident. In every single case, it results from a systematic survey of innovative opportunities. In every single case, the entrepreneur looks for the place where a specialty skill can be developed and can give a new enterprise a unique controlling position. Robert Bosch spent years studying the new automotive field to position his new company where it could immediately establish itself as the leader. Hamilton Propeller, for many years the leading airplane propeller manufacturer in the United States, was the result of a systematic search by its founder in the early days of powered flight. Baedeker made several attempts to start a service for the tourist before he decided on the guidebook that then bore his name and made him famous.

The first point, therefore, is that in the early stages of a new industry, a new market, or a new major trend, there is the opportunity to search systematically for the specialty skill opportunity—and then there is usually time to develop a unique skill.

The second point is that the specialty skill niche does require a skill that is both unique and different. The early automobile pioneers were, without exception, mechanics. They knew a great deal about machinery, about metals and about engines. But electricity was alien to them.

It required theoretical knowledge which they neither possessed nor knew how to acquire. There were other publishers in Baedeker's time, but a guidebook that required on-the-spot gathering of an enormous amount of detailed information, constant inspection, and a staff of traveling auditors was not within their purview. "Counter-trade" is neither trading nor banking.

The business that establishes itself in a specialty skill niche is therefore unlikely to be threatened by its customers or by its suppliers. Neither of them really wants to get into something that is so alien in skill and in temperament.

Thirdly, a business occupying a specialty skill niche must constantly work on improving its own skill. It has to stay ahead. Indeed, it has to make itself constantly obsolete. The automobile companies in the early days used to complain that Delco in Dayton, and Bosch in Stuttgart, were pushing them. They turned out lighting systems that were far ahead of the ordinary automobile, ahead of what the automobile manufacturers of the times thought the customer needed, wanted, or could pay for, ahead very often of what the automobile manufacturer knew how to assemble.

While the specialty skill niche has unique advantages, it also has severe limitations. One is that it inflicts tunnel-vision on its occupants. In order to maintain themselves in their controlling position, they have to learn to look neither right nor left, but directly ahead at their narrow area, their specialized field. Airplane electronics were not too different from automobile electronics in the early stages. Yet the automobile electricians—Delco, Bosch, and Lucas—are not leaders in airplane electronics. They did not even see the field and made no attempt to get into it.

A second, serious limitation is that the occupant of a specialty skill niche is usually dependent on somebody else to bring his product or service to market. It becomes a component. The strength of the automobile electrical firms is that the customer does not know that they exist. But this is of course also their weakness. If the British automobile industry goes down, so does Lucas. A. O. Smith prospered making automotive frames until the energy crisis. Then American automobile manufacturers began to switch to cars without frames. These cars are substantially more expensive than cars with frames, but they weigh less and therefore burn less fuel. A. O. Smith could do nothing to reverse the adverse trend.

Finally, the greatest danger to the specialty niche manufacturer is for the specialty to cease being a specialty and to become universal.

The niche that the Viennese counter-traders now occupy was occupied in the 1920s and 1930s by foreign exchange traders who were mostly Swiss. Bankers of those days, having grown up before World War I, still believed that currencies ought to be stable. And when currencies became unstable, when there were blocked currencies around, currencies with different exchange rates for different purposes, and other such monstrosities, the bankers did not even want to handle the business. They were only too happy to let the specialists in Switzerland do what they thought was a dirty job. So a fairly small number of Swiss foreign exchange traders occupied a highly profitable specialty skill niche. After World War II, with the tremendous expansion of world trade, foreign exchange trading became routine. By now every bank, at least in the major money centers, has its own foreign exchange traders.

The specialty skill niche, like all ecological niches, is therefore limited—in scope as well as in time. Species that occupy such a niche, biology teaches, do not easily adapt to even small changes in the external environment. And this is true, too, of the entrepreneurial skill species. But within these limitations, the specialty skill niche is a highly advantageous position. In a rapidly expanding new technology, industry, or market, it is perhaps the most advantageous strategy. Very few of the automobile makers of 1920 are still around; every single one of the electrical and lighting systems makers is. Once attained and properly maintained, the specialty skill niche protects against competition, precisely because no automobile buyer knows or cares who makes the headlights or the brakes. No automobile buyer is therefore likely to shop around for either. Once the name "Baedeker" had become synonymous with tourist guidebooks, there was little danger that anybody else would try to muscle in, at least not until the market changed drastically. In a new technology, a new industry, or a new market, the specialty skill strategy offers an optimal ratio between opportunity and risk of failure.

III

THE SPECIALTY MARKET

The major difference between the specialty skill niche and the specialty market niche is that the former is built around a product or service and the latter around specialized knowledge of a market. Otherwise, they are similar.

Two medium-sized companies, one in northern England and one in Denmark, supply the great majority of the automated baking ovens for cookies and crackers bought in the non-Communist world. For many decades, two companies—the two earliest travel agents, Thomas Cook in Europe and American Express in the United States—had a practical monopoly on travelers checks.

There is, I am told, nothing very difficult or particularly technical about baking ovens. There are literally dozens of companies around that could make them just as well as those two firms in England and Denmark. But these two know the market: they know every single major baker, and every single major baker knows them. The market is just not big enough or attractive enough to try to compete with these two, as long as they remain satisfactory. Similarly, travelers checks were a backwater until the post–World War II period of mass travel. They were highly profitable since the issuer, whether Cook or American Express, has the use of the money and keeps the interest earned on it until the purchaser cashes the check—sometimes months after the checks were purchased. But the market was not large enough to tempt anyone else. Furthermore, travelers checks required a worldwide organization, which Cook and American Express had to maintain anyhow to service their travel customers, and which nobody else in those days had any reason to build.

The specialty market is found by looking at a new development with the question, What opportunities are there in this that would give us a unique niche, and what do we have to do to fill it ahead of everybody else? The travelers check is no great "invention." It is basically nothing more than a letter of credit, and that has been around for hundreds of years. What was new was that travelers checks were offered—at first to the customers of Cook and American Express, and then to the general public—in standard denominations. And they could be cashed wherever Cook or American Express had an office or an agent. That made them uniquely attractive to the tourist who did not want to carry a great deal of cash and did not have the established banking connections to make them eligible for a letter of credit.

There was nothing particularly advanced in the early baking ovens, nor is there any high technology in the baking ovens installed today. What the two leading firms did was to realize that the act of baking cookies and crackers was moving out of the home and into the factory. They then studied what commercial bakers needed so that they could manufacture the product their own customers, grocers and supermarkets, could in turn sell and the housewife would buy. The baking ovens were not based on engineering but on market research; the engineering would have been available to anyone.

The specialty market niche has the same requirements as the specialty skill niche: systematic analysis of a new trend, industry, or market; a specific innovative contribution, if only a "twist" like the one that converted the traditional letter of credit into the modern travelers check; and continuous work to improve the product and especially the service, so that leadership, once obtained, will be retained.

And it has the same limitations. The greatest threat to the specialty market position is success. The greatest threat is when the specialty market becomes a mass market.

Travelers checks have now become a commodity and highly competitive because travel has become a mass market.

So have perfumes. A French firm, Coty, created the modern perfume industry. It realized that World War I had changed the attitude toward cosmetics. Whereas before the war only "fast women" used cosmetics—or dared admit to their use—cosmetics had become accepted and respectable. By the mid-twenties Coty had established itself in what was almost a monopoly position on both sides of the Atlantic. Until 1929 the cosmetics market was a "specialty market," a market of the upper middle class. But then during the Depression it exploded into a genuine mass market. It also split into two segments: a prestige segment, with high prices, specialty distribution, and specialty packaging; and popular-priced, mass brands sold in every outlet including the supermarket, the variety store, and the drugstore. Within a few short years, the specialty market dominated by Coty had disappeared. But Coty could not make up its mind whether to try to become one of the mass marketers in cosmetics or one of the luxury producers. It tried to stay in a market that no longer existed, and has been drifting ever since.

19

Changing Values and Characteristics

In the entrepreneurial strategies discussed so far, the aim is to introduce an innovation. In the entrepreneurial strategy discussed in this
chapter, the strategy itself is the innovation. The product or service it
carries may well have been around a long time—in our first example,
the postal service, it was almost two thousand years old. But the strategy converts this old, established product or service into something
new. It changes its utility, its value, its economic characteristics. While
physically there is no change, economically there is something different and new.

All the strategies to be discussed in this chapter have one thing in
common. They create a customer—and that is the ultimate purpose of
a business, indeed, of economic activity.* But they do so in four different ways:

- by creating utility;
- by pricing;
- by adaptation to the customer's social and economic reality;
- by delivering what represents true value to the customer.

I

CREATING CUSTOMER UTILITY

English schoolboys used to be taught that Rowland Hill "invented"
the postal service in 1836. That is nonsense, of course. The Rome of the
Caesars had an excellent service, with fast couriers carrying mail on
regular schedules to the furthest corners of the Empire. A thousand
years later, in 1521, the German emperor Charles V, in true Renais-

*As was first said more than thirty years ago in my *The Practice of Management* (New
York: Harper & Row, 1954).

sance fashion, went back to Classical Rome and gave a monopoly on carrying mail in the imperial domains to the princely family of Thurn and Taxis. Their generous campaign contributions had enabled him to bribe enough German Electors to win the imperial crown—and the princes of Thurn and Taxis still provided the postal service in many parts of Germany as late as 1866, as stamp collectors know. By the middle of the seventeenth century, every European country had organized a postal service on the German model and so had, a hundred years later, the American colonies. Indeed, all the great letter-writers of the Western tradition, from Cicero to Madame de Sévigné, Lord Chesterfield, and Voltaire, wrote and posted their letters long before Rowland Hill "invented" the postal service.

Yet Hill did indeed create what we would now call "mail." He contributed no new technology and not one new "thing," nothing that could conceivably have been patented. But mail had always been paid for by the addressee, with the fee computed according to distance and weight. This made it both expensive and slow. Every letter had to be brought to a post office to be weighed. Hill proposed that postage should be uniform within Great Britain regardless of distance; that it be prepaid; and that the fee be paid by affixing the kind of stamp that had been used for many years to pay other fees and taxes. Overnight, mail became easy and convenient; indeed, letters could now be dropped into a collection box. Immediately, also, mail became absurdly cheap. The letter that had earlier cost a shilling or more—and a shilling was as much as a craftsman earned in a day—now cost only a penny. The volume was no longer limited. In short, "mail" was born.

Hill created utility. He asked: What do the customers *need* for a postal service to be truly a service to them? This is always the first question in the entrepreneurial strategy of changing utility, values, and economic characteristics. In fact, the reduction in the cost of mailing a letter, although 80 percent or more, was secondary. The main effect was to make using the mails convenient for everybody and available to everybody. Letters no longer had to be confined to "epistles." The tailor could now use the mail to send a bill. The resulting explosion in volume, which doubled in the first four years and quadrupled again in the next ten, then brought the cost down to where mailing a letter cost practically nothing for long years.

Price is usually almost irrelevant in the strategy of creating utility. The strategy works by enabling customers to do what serves *their pur-*

pose. It works because it asks: What is truly a "service," truly a "utility" to the customer?

Every American bride wants to get one set of "good china." A whole set is, however, far too expensive a present, and the people giving her a wedding present do not know what pattern the bride wants or what pieces she already has. So they end up giving something else. The demand was there, in other words, but the utility was lacking. A medium-sized dinnerware manufacturer, the Lenox China Company, saw this as an innovative opportunity. Lenox adapted an old idea, the "bridal register," so that it only "registers" Lenox china. The bride-to-be then picks one merchant whom she tells what pattern of Lenox china she wants, and to whom she refers potential donors of wedding gifts. The merchant then asks the donor: "How much do you want to spend?" and explains: "That will get you two coffee cups with saucers." Or the merchant can say, "She already has all the coffee cups; what she needs now is dessert plates." The result is a happy bride, a happy wedding-gift donor, and a very happy Lenox China Company.

Again, there is no high technology here, nothing patentable, nothing but a focus on the needs of the customer. Yet the bridal register, for all its simplicity—or perhaps because of it—has made Lenox the favorite "good china" manufacturer and one of the most rapidly growing of medium-sized American manufacturing companies.

Creating utility enables people to satisfy their wants and their needs *in their own way.* The tailor could not send the bill to his customer through the mails if it first took three hours to get the letter accepted by a postal clerk and if the addressee then had to pay a large sum—perhaps even as much as the bill itself. Rowland Hill did not add anything to the service. It was performed by the same postal clerks using the same mail coaches and the same letter carriers. And yet Rowland Hill's postal service was a totally different "service." It served a different function.

II

PRICING

For many years, the best known American face in the world was that of King Gillette, which graced the wrapper of every Gillette razor blade

246   ENTREPRENEURIAL STRATEGIES

sold anyplace in the world. And millions of men all over the world used
a Gillette razor blade every morning.

King Gillette did not invent the safety razor; dozens of them were
patented in the closing decades of the nineteenth century. Until 1860
or 1870, only a very small number of men, the aristocracy and a few
professionals and merchants, had to take care of their facial hair, and
they could well afford a barber. Then, suddenly, large numbers of men,
tradesmen, shopkeepers, clerks, had to look "respectable." Few of them
could handle a straight razor or felt comfortable with so dangerous a
tool, but visits to the barber were expensive, and worse, time-consum-
ing. Many inventors designed a "do-it-yourself" safety razor, yet none
could sell it. A visit to the barber cost ten cents and the cheapest safety
razor cost five dollars—an enormous sum in those days when a dollar
a day was a good wage.

Gillette's safety razor was no better than many others, and it was a
good deal more expensive to produce. But Gillette did not "sell" the
razor. He practically gave it away by pricing it at fifty-five cents retail
or twenty cents wholesale, not much more than one-fifth of its manufacturing cost. But he designed it so that it could use only his patented
blades. These cost him less than one cent apiece to make: he sold them
for five cents. And since the blades could be used six or seven times,
they delivered a shave at less than one cent apiece—or at less than
one-tenth the cost of a visit to a barber.

What Gillette did was to price what the customer buys, namely, the
shave, rather than what the manufacturer sells. In the end, the captive
Gillette customer may have paid more than he would have paid had he
bought a competitor's safety razor for five dollars, and then bought the
competitor's blades selling at one cent or two. Gillette's customers
surely knew this; customers are more intelligent than either advertising
agencies or Ralph Nader believe. But Gillette's pricing made sense to
them. They were paying for what they bought, that is, for a shave,
rather than for a "thing." And the shave they got from the Gillette razor
and the Gillette razor blade was much more pleasant than any shave
they could have given themselves with that dangerous weapon, the
straight-edge razor, and far cheaper than they could have gotten at the
neighborhood barber's.

One reason why the patents on a copying machine ended up at a
small, obscure company in Rochester, New York, then known as the
Haloid Company, rather than at one of the big printing-machine manu-

facturers, was that none of the large established manufacturers saw any possibility of selling a copying machine. Their calculations showed that such a machine would have to sell for at least $4,000. Nobody was going to pay such a sum for a copying machine when carbon paper cost practically nothing. Also, of course, to spend $4,000 on a machine meant a capital-appropriations request, which had to go all the way up to the board of directors accompanied by a calculation showing the return on investment, both of which seemed unimaginable for a gadget to help the secretary. The Haloid Company—the present Xerox—did a good deal of technical work to design the final machine. But its major contribution was in pricing. It did not sell the machine; it sold what the machine produced, copies. At five or ten cents a copy, there is no need for a capital-appropriations request. This is "petty cash," which the secretary can disburse without going upstairs. Pricing the Xerox machine at five cents a copy was the true innovation.

Most suppliers, including public-service institutions, never think of pricing as a strategy. Yet pricing enables the customer to pay for what he buys—a shave, a copy of a document—rather than for what the supplier makes. What is being paid in the end is, of course, the same amount. But how it is being paid is structured to the needs and the realities of the consumer. It is structured in accordance with what the consumer actually buys. And it charges for what represents "value" to the customer rather than what represents "cost" to the supplier.

III

THE CUSTOMER'S REALITY

The worldwide leadership of the American General Electric Company (G.E.) in large steam turbines is based on G.E.'s having thought through, in the years before World War I, what its customers' realities were. Steam turbines, unlike the piston-driven steam engines which they replaced in the generation of electric power, are complex, requiring a high degree of engineering in their design, and skill in building and fitting them. This the individual electric power company simply cannot supply. It buys a major steam turbine maybe every five or ten years when it builds a new power station. Yet the skill has to be kept in being all the time. The manufacturer, therefore, has to set up and maintain a massive consulting organization.

But, as G.E. soon found out, the customer cannot pay for consulting services. Under American law, the state public utility commissions would have to allow such an expenditure. In the opinion of the commissions, however, the companies should have been able to do this work themselves. G.E. also found that it could not add to the price of the steam turbine the cost of the consulting services which its customers needed. Again, the public utility commissions would not have accepted it. But while a steam turbine has a very long life, it needs a new set of blades fairly often, maybe every five to seven years, and these blades have to come from the maker of the original turbine. G.E. built up the world's foremost consulting engineering organization on electric power stations—though it was careful not to call this consulting engineering but "apparatus sales"—for which it did not charge. Its steam turbines were no more expensive than those of its competitors. But it put the added cost of the consulting organization plus a substantial profit into the price it charged for replacement blades. Within ten years all the other manufacturers of steam turbines had caught on and switched to the same system. But by then G.E. had world market leadership.

Much earlier, during the 1840s, a similar design of product and process to fit customer realities led to the invention of installment buying. Cyrus McCormick was one of many Americans who built a harvesting machine—the need was obvious. And he found, as had the other inventors of similar machines, that he could not sell his product. The farmer did not have the purchasing power. That the machine would earn back what it cost within two or three seasons, everybody knew and accepted, but there was no banker then who would have lent the American farmer the money to buy a machine. McCormick offered installments, to be paid out of the savings the harvester produced over the ensuing three years. The farmer could now afford to buy the machine—and he did so.

Manufacturers are wont to talk of the "irrational customer" (as do economists, psychologists, and moralists). But there are no "irrational customers." As an old saying has it, "There are only lazy manufacturers." The customer has to be assumed to be rational. His or her reality, however, is usually quite different from that of the manufacturer. The rules and regulations of public utility commissions may appear to make no sense and be purely arbitrary. For the power companies that have to operate under them, they are realities nonetheless. The American farmer may have been a better credit risk than American bankers of

1840 thought. But it was a fact that American banks of that period did not advance money to farmers to purchase equipment. The innovative strategy consists in accepting that these realities are not extraneous to the product, but *are*, in fact, the product as far as the customer is concerned. Whatever customers buy has to fit their realities, or it is of no use to them.

IV

DELIVERING VALUE TO THE CUSTOMER

The last of these innovative strategies delivers what is "value" to the customer rather than what is "product" to the manufacturer. It is actually only one step beyond the strategy of accepting the customer's reality as part of the product and part of what the customer buys and pays for.

A medium-sized company in America's Midwest supplies more than half of all the special lubricant needed for very large earth-moving and hauling machines: the bulldozers and draglines used by contractors building highways; the heavy equipment used to remove the overlay from strip mines; the heavy trucks used to haul coal out of coal mines; and so on. This company is in competition with some of the largest oil companies, which can mobilize whole battalions of lubrication specialists. It competes by not selling lubricating oil at all. Instead, it sells what is, in effect, insurance. What is "value" to the contractor is not lubrication: it is operating the equipment. Every hour the contractor loses because this or that piece of heavy equipment cannot operate costs him infinitely more than he spends on lubricants during an entire year. In all these activities there is a heavy penalty for contractors who miss their deadlines—and they can only get the contract by calculating the deadline as finely as possible and racing against the clock. What the Midwestern lubricant maker does is to offer contractors an analysis of the maintenance needs of their equipment. Then it offers them a maintenance program with an annual subscription price, and guarantees the subscribers that their heavy equipment will not be shut down for more than a given number of hours per year because of lubrication problems. Needless to say, the program always prescribes the manufacturer's lubricant. But this is not what contractors buy. They are buying trouble-free operations, which are extremely valuable to them.

The final example—one that might be called "moving from product to system"—is that of Herman Miller, the American furniture maker in Zeeland, Michigan. The company first became well known as the manufacturer of one of the early modern designs, the Eames chair. Then, when every other manufacturer began to turn out designer chairs, Herman Miller moved into making and selling whole offices and work stations for hospitals, both with considerable success. Finally, when the "office of the future" began to come in, Herman Miller founded a Facilities Management Institute that does not even sell furniture or equipment, but advises companies on office layout and equipment needed for the best work flow, high productivity, high employee morale, all at low cost. What Herman Miller is doing is *defining* "value" for the customer. It is telling the customer, "You may pay for furniture, but you are buying work, morale, productivity. And this is what you should therefore be paying for."

These examples are likely to be considered obvious. Surely, anybody applying a little intelligence would have come up with these and similar strategies? But the father of systematic economics, David Ricardo, is believed to have said once, "Profits are not made by differential cleverness, but by differential stupidity." The strategies work, not because they are clever, but because most suppliers—of goods as well as of services, businesses as well as public-service institutions—do not think. They work precisely because they are so "obvious." Why, then, are they so rare? For, as these examples show, anyone who asks the question, What does the customer really buy? will win the race. In fact, it is not even a race since nobody else is running. What explains this?

One reason is the economists and their concept of "value." Every economics book points out that customers do not buy a "product," but what the product does for them. And then, every economics book promptly drops consideration of everything except the "price" for the product, a "price" defined as what the customer pays to take possession or ownership of a thing or a service. What the product does for the customer is never mentioned again. Unfortunately, suppliers, whether of products or of services, tend to follow the economists.

It is meaningful to say that "product A costs X dollars." It is meaningful to say that "we have to get Y dollars for the product to cover our own costs of production and have enough left over to cover the cost of capital, and thereby to show an adequate profit." But it makes no sense at all to conclude, " . . . and therefore the customer has to pay the lump sum of Y dollars in cash for each piece of product A he buys." Rather, the argument should go as follows: "What the customer pays for each piece of the product has to work out as Y dollars *for us*. But how the customer pays depends on what makes the most sense to him. It depends on what the product does for the customer. It depends on what fits his reality. It depends on what the customer sees as 'value.' "

Price in itself is not "pricing," and it is not "value." It was this insight that gave King Gillette a virtual monopoly on the shaving market for almost forty years; it also enabled the tiny Haloid Company to become the multibillion-dollar Xerox Company in ten years, and it gave General Electric world leadership in steam turbines. In every single case, these companies became exceedingly profitable. But they earned their profitability. They were paid for giving their customers satisfaction, for giving their customers what the customers wanted to buy, in other words, for giving their customers their money's worth.

"But this is nothing but elementary marketing," most readers will protest, and they are right. It is *nothing* but elementary marketing. To start out with the customer's utility, with what the customer buys, with what the realities of the customer are and what the customer's values are—this is what marketing is all about. But why, after forty years of preaching Marketing, teaching Marketing, professing Marketing, so few suppliers are willing to follow, I cannot explain. The fact remains that so far, anyone who is willing to use marketing as the basis for strategy is likely to acquire leadership in an industry or a market fast and almost without risk.

Entrepreneurial strategies are as important as purposeful innovation and entrepreneurial management. Together, the three make up *innovation and entrepreneurship.*

The available strategies are reasonably clear, and there are only a few of them. But it is far less easy to be specific about entrepreneurial strategies than it is about purposeful innovation and entrepreneurial management. We know what the areas are in which innovative opportunities are to be found and how they are to be analyzed. There are correct policies and practices and wrong policies and practices to make an existing business or public-service institution capable of entrepreneurship; right things to do and wrong things to do in a new venture. But the entrepreneurial strategy that fits a certain innovation is a high-risk decision. Some entrepreneurial strategies are better fits in a given situation, for example, the strategy that I called entrepreneurial judo, which is the strategy of choice where the leading businesses in an industry persist year in and year out in the same habits of arrogance and false superiority. We can describe the typical advantages and the typical limitations of certain entrepreneurial strategies.

Above all, we know that an entrepreneurial strategy has more chance of success the more it starts out with the users—their utilities, their values, their realities. An innovation is a change in market or society. It produces a greater yield for the user, greater wealth-producing capacity for society, higher value or greater satisfaction. The test of an innovation is always what it does for the user. Hence, entrepreneurship always needs to be market-focused, indeed, market-driven.

Still, entrepreneurial strategy remains the decision-making area of entrepreneurship and therefore the risk-taking one. It is by no means hunch or gamble. But it also is not precisely science. Rather, it is judgment.

Conclusion:
The Entrepreneurial Society

I

"Every generation needs a new revolution," was Thomas Jefferson's conclusion toward the end of his long life. His contemporary, Goethe, the great German poet, though an arch-conservative, voiced the same sentiment when he sang in his old age:

Vernunft wird Unsinn
Wohltat, Plage. *

Both Jefferson and Goethe were expressing their generation's disenchantment with the legacy of Enlightenment and French Revolution. But they might just as well have reflected on our present-day legacy, 150 years later, of that great shining promise, the Welfare State, begun in Imperial Germany for the truly indigent and disabled, which has now become "everybody's entitlement" and an increasing burden on those who produce. Institutions, systems, policies eventually outlive themselves, as do products, processes, and services. They do it when they accomplish their objectives and they do it when they fail to accomplish their objectives. The mechanisms may still tick. But the assumptions on which they were designed have become invalid—as, for example, have the demographic assumptions on which health-care plans and retirement schemes were designed in all developed countries over the last hundred years. Then, indeed, reason becomes nonsense and boons afflictions.

Yet "revolutions," as we have learned since Jefferson's days, are not the remedy. They cannot be predicted, directed, or controlled. They

*Reason becomes nonsense, /Boons afflictions.

253

bring to power the wrong people. Worst of all, their results—predictably—are the exact opposite of their promises. Only a few years after Jefferson's death in 1826, that great anatomist of government and politics, Alexis de Tocqueville, pointed out that revolutions do not demolish the prisons of the old regime, they enlarge them. The most lasting legacy of the French Revolution, Tocqueville proved, was the tightening of the very fetters of pre-Revolutionary France: the subjection of the whole country to an uncontrolled and uncontrollable bureaucracy, and the centralization in Paris of all political, intellectual, artistic, and economic life. The main consequences of the Russian Revolution were new serfdom for the tillers of the land, an omnipotent secret police, and a rigid, corrupt, stifling bureaucracy—the very features of the czarist regime against which Russian liberals and revolutionaries had protested most loudly and with most justification. And the same must be said of Mao's macabre "Great Cultural Revolution."

Indeed, we now know that "revolution" is a delusion, the pervasive delusion of the nineteenth century, but today perhaps the most discredited of its myths. We now know that "revolution" is not achievement and the new dawn. It results from senile decay, from the bankruptcy of ideas and institutions, from failure of self-renewal.

And yet we also know that theories, values, and all the artifacts of human minds and human hands do age and rigidify, becoming obsolete, becoming "afflictions."

Innovation and entrepreneurship are thus needed in society as much as in the economy, in public-service institutions as much as in businesses. It is precisely because innovation and entrepreneurship are not "root and branch" but "one step at a time," a product here, a policy there, a public service yonder; because they are not planned but focused on this opportunity and that need; because they are tentative and will disappear if they do not produce the expected and needed results; because, in other words, they are pragmatic rather than dogmatic and modest rather than grandiose—that they promise to keep any society, economy, industry, public service, or business flexible and self-renewing. They achieve what Jefferson hoped to achieve through revolution in every generation, and they do so without bloodshed, civil war, or concentration camps, without economic catastrophe, but with purpose, with direction, and under control.

What we need is an entrepreneurial society in which innovation and entrepreneurship are normal, steady, and continuous. Just as management has become the specific organ of all contemporary institutions, and the integrating organ of our society of organizations, so innovation and entrepreneurship have to become an integral life-sustaining activity in our organizations, our economy, our society.

This requires of executives in all institutions that they make innovation and entrepreneurship a normal, ongoing, everyday activity, a practice in their own work and in that of their organization. To provide concepts and tools for this task is the purpose of this book.

II

WHAT WILL NOT WORK

The first priority in talking about the public policies and governmental measures needed in the entrepreneurial society is to define what will not work—especially as the policies that will not work are so popular today.

"Planning" as the term is commonly understood is actually incompatible with an entrepreneurial society and economy. Innovation does indeed need to be purposeful and entrepreneurship has to be managed. But innovation, almost by definition, has to be decentralized, *ad hoc,* autonomous, specific, and micro-economic. It had better start small, tentative, flexible. Indeed, the opportunities for innovation are found, on the whole, only way down and close to events. They are not to be found in the massive aggregates with which the planner deals of necessity, but in the deviations therefrom—in the unexpected, in the incongruity, in the difference between "The glass is half full" and "The glass is half empty," in the weak link in a process. By the time the deviation becomes "statistically significant" and thereby visible to the planner, it is too late. Innovative opportunities do not come with the tempest but with the rustling of the breeze.

It is popular today, especially in Europe, to believe that a country can have "high-tech entrepreneurship" by itself. France, West Germany, even England are basing national policies on this premise. But it is a delusion. Indeed, a policy that promotes high tech and high tech alone—and that otherwise is as hostile to entrepreneurship as France, West Germany, and even England still are—will not even produce high tech. All it can come up with is another expensive flop, another supersonic *Concorde*; a little *gloire,* oceans of red ink, but neither jobs nor technological leadership.

High tech in the first place—and this is, of course, one of the major premises of this book—is only one area of innovation and entrepreneurship. The great bulk of innovations lies in other areas. But also, a high-tech policy will run into political obstacles that will defeat it in short order. In terms of job creation, high tech is the maker of tomorrow rather than the maker of today. As we saw initially (in the Introduction), "high tech" in the United States created no more jobs in the period 1970–85 than "smokestack" lost: about five to six million. *All* the additional jobs in the American economy during that period—a total of 35 million—were created by new ventures that were not "high-tech" but "middle-tech," "low-tech," or "no-tech." The European countries, however, will be under increasing pressure to find additional jobs for a growing work force. And if then the focus in innovation and entrepreneurship is high-tech, the demand that governments abandon the high-tech policies which sacrifice the needs of today—the bolstering of the ailing industrial giants—to the uncertain promise of a high-tech future will become irresistible. In France this has been the issue over which the Communists pulled out of President Mitterand's cabinet in 1984, and the left wing of Mitterand's own Socialist Party is also increasingly unhappy and restless.

Above all, to have "high-tech" entrepreneurship alone without its being embedded in a broad entrepreneurial economy of "no-tech," "low-tech," and "middle-tech," is like having a mountaintop without the mountain. Even high-tech people in such a situation will not take jobs in new, risky, high-tech ventures. They will prefer the security of a job in the large, established, "safe" company or in a government agency. Of course, high-tech ventures need a great many people who are not themselves high-tech: accountants, salespeople, managers, and so on. In an economy that spurns entrepreneurship and innovation except for that tiny extravaganza, the "glamorous high-tech venture," those people will keep on looking for jobs and career opportunities where society and economy (i.e., their classmates, their parents, and their teachers) encourage them to look: in the large, "safe," established institution. Neither will distributors be willing to take on the products of the new venture, nor investors be willing to back it.

But the other innovative ventures are also needed to supply the capital that high tech requires. Knowledge-based innovation, and in particular high-tech innovation, has the longest lead time between investment and profitability. The world's computer industry did not break even until the late seventies, that is, after thirty loss years. To be sure, IBM made very good money quite early. And one after another of the "Seven Dwarfs," the smaller American computer makers, moved into the black during the late sixties. But these profits were offset several times over by the tremendous losses of all the others, and especially of the big old companies who failed totally in computers: General Electric, Westinghouse, ITT, and RCA in America; the (British) General Electric Company, Ferranti, and Plessey in Great Britain; Thomson-Houston in France; Siemens and Telefunken in Germany; Philips in Holland; and many others. History is repeating itself now in minicomputers and personal computers: it will be many years before the industry worldwide moves into the black. And the same thing is happening in biotechnology. This was also the pattern a hundred years ago in the electrical apparatus industry of the 1880s, for instance, or in the automobile industry of 1900 or 1910.

And during this long gestation period, non-high-tech ventures have to produce the profits to offset the losses of high tech and provide the needed capital.

The French are right, of course: economic and political strength these days requires a high-tech position, whether in information technology, in biology, or in automation. The French surely have the scientific and technical capacity. And yet it is most unlikely (I am tempted to say impossible) for any country to be innovative and entrepreneurial in high tech without having an entrepreneurial economy. High tech is indeed the leading edge, but there cannot be an edge without a knife. There cannot be a viable high-tech sector by itself any more than there can be a healthy brain in a dead body. There must be an economy full of innovators and entrepreneurs, with entrepreneurial vision and entrepreneurial values, with access to venture capital, and filled with entrepreneurial vigor.

III

THE SOCIAL INNOVATIONS NEEDED

There are two areas in which an entrepreneurial society requires substantial social innovation.

1. The first is a policy to take care of redundant workers. The numbers are not large. But blue-collar workers in "smokestack industries" are concentrated in a very few places; three-quarters of all American

automobile workers live in twenty counties, for instance. They are therefore highly visible, and they are highly organized. More important, they are ill equipped to place themselves, to redirect themselves, to move. They have neither education nor skill nor social competence —and above all not much self-confidence. They never applied for a job throughout their life; when they were ready to go to work, a relative already working in the automobile plant introduced them to the supervisor. Or the parish priest gave them a letter to one of his parishioners who was already working in the mill. And the "smokestack" workers in Great Britain—or the Welsh coal miners—are no different, nor are the blue-collar workers in Germany's Ruhr, in Lorraine, or in the Belgian Borinage. These workers are the one group in developed societies that have not experienced in this century a tremendous growth in education and horizon. In respect to competence, experience, skill, and schooling they are pretty much where the unskilled laborer of 1900 was. The one thing that has happened to them is an explosive rise in their incomes —on balance they are the highest-paid group in industrial society if wages and benefits are added together—and in political power as well. They therefore do not have enough capacity, whether as individuals or as a group, to help themselves, but more than enough power to oppose, to veto, to impede. Unless society takes care of placing them—if only in lower-paying jobs—they must become a purely negative force.

The problem is soluble if an economy becomes entrepreneurial. For then the new businesses of the entrepreneurial economy create new jobs, as has been happening in the United States during the last ten years (which explains why the massive unemployment in the old "smokestack industries" has caused so little political trouble so far in the United States and has not even triggered a massive protectionist reaction). But even if an entrepreneurial economy creates the new jobs, there is need for organized efforts to train and place the redundant former "smokestack" workers—they cannot do it by themselves. Otherwise redundant "smokestack" labor will increasingly oppose anything new, including even the means of their own salvation. The "mini-mill" offers jobs to redundant steel workers. The automated automobile plant is the most appropriate work place for displaced automobile workers. And yet both the "mini-mill" and automation in the car factory are bitterly fought by the present workers—even though they know that their own jobs will not last. Unless we can make innovation an opportunity for redundant workers in the "smokestack" industries their feeling of impotence, their fears, their sense of being caught will lead them to resist all innovation—as is already the case in Great Britain (or in the U.S. Postal Service). The job has been done before—by the Mitsui *Zaibatsu* of Japan in the sharp Japanese Depression after the Russo-Japanese war of 1906, by the Swedes after World War II in the deliberate policy which converted a country of subsistence farmers and forest workers into an industrialized and highly prosperous nation. And the numbers are, as already said, not very large—especially as we need not concern ourselves overmuch with the one-third of the group that is fifty-five years old and older and has available adequate early-retirement provisions, and with another third that is under thirty years of age and capable of moving and of placing themselves. But the policy to train and place the remaining one-third—a small but hard core—of displaced "smokestack" workers has yet to be worked out.

2. The other social innovation needed is both more radical and more difficult and unprecedented: to organize the systematic abandonment of outworn social policies and obsolete public-service institutions. This was not a problem in the last great entrepreneurial era; a hundred years ago there were few such policies and institutions. Now we have them in abundance. But by now we also know that few if any are for ever. Few of them even perform more than a fairly short time.

One of the fundamental changes in world view and perception of the last twenty years—a truly monumental turn—is the realization that governmental policies and agencies are of human rather than of divine origin, and that therefore the one thing certain about them is that they will become obsolete fairly fast. Yet politics is still based on the age-old assumption that whatever government does is grounded in the nature of human society and therefore "forever." As a result there is no political mechanism so far to slough off the old, the outworn, the no-longer-productive in government.

Or rather what we have is not working yet. In the United States there has lately been a rash of "sunset laws," which prescribe that a governmental agency or a public law lapse after a certain period of time unless specifically re-enacted. These laws have not worked, however—in part because there are no objective criteria as to when an agency or a law becomes dysfunctional; in part because there is so far no organized process of abandonment; but perhaps mostly because we have not yet learned to develop new or alternative methods for achieving what an ineffectual law or agency was originally supposed to achieve.

To develop both the principles and the process for making "sunset laws" meaningful and effective is one of the important social innovations ahead of us—and one that needs to be made soon. Our societies are ready for it.

I V

THE NEW TASKS

These two social policies needed are, however, only examples. Underlying them is the need for a massive reorientation in policies and attitudes, and above all, in priorities. We need to encourage habits of flexibility, of continuous learning, and of acceptance of change as normal and as opportunity—for institutions as well as for individuals.

Tax policy is one area—important both for its impact on behavior and as a symbol of society's values and priorities. In developed countries, sloughing off yesterday is at present severely penalized by the tax system. In the United States, for instance, the tax collector treats monies realized by selling or liquidating a business or a product line as income. Actually the amounts are, of course, repayments of capital. But under the present tax system the company pays corporation income tax on them. And if it distributes the proceeds to its shareholders, they pay full personal income tax on them as if they were ordinary "dividends"—that is, distribution of "profits." As a result businesses prefer not to abandon the old, the obsolescent, the no-longer-productive; they'd rather hang on to it and keep on pouring money into it. Worse still, they then assign their most capable people to "defending" the outworn in a massive misallocation of the scarcest and most valuable resource—the human resource that needs to be allocated to making tomorrow, if the company is to have a tomorrow. And when the company then finally liquidates or sells the old, obsolescent, no-longer-productive business or product line, it does not distribute the proceeds to the shareholders and does not therefore return them to the capital market where they become available for investment in innovative entrepreneurial opportunities. Rather the company keeps these funds and commonly invests them in its old, traditional, declining business or products—that is, into those parts of its operations and activities for which it could not easily raise money on the capital market—again resulting in a massive misallocation of scarce resources.

What is needed in an entrepreneurial society is a tax system that encourages moving capital from yesterday into tomorrow rather than one that, like our present one, prevents and penalizes it.

But we also should be able in and through the tax system to assuage the most pressing financial problem of the new and growing business: cash shortage. One way might be acceptance of economic reality: during the first five or six years of the life of a new, and particularly of a growing, business, "profits" are an accounting fiction. During these years the costs of staying in business are always—and almost by definition—larger for a new venture than the surplus from yesterday's operations (that is, the excess of current income over yesterday's costs). This means in effect that a new and growing venture always has to invest every penny of operating surplus to stay alive; usually, especially if growing fast, it has to invest a good deal more than it can possibly hope to produce as "current surplus" (that is, as "profit") in its current accounts. For the first few years of its life the new and growing venture —whether standing by itself or part of an existing enterprise—should therefore be exempt from income taxes, for the same reason for which we do not expect a small and rapidly growing child to produce a "surplus" that supports a grown-up. And taxes are the means by which a producer supports somebody else—namely, a nonproducer. By the way, exempting the new venture from taxation until it has "grown up" would almost certainly in the end produce a substantially higher tax yield.

If this, however, is deemed too "radical," the new venture should at least be able to postpone paying taxes on the so-called profits of its infant years. It should be able to retain the cash until it is past the period of acute cash-flow pressure, and to do so without penalty or interest charges.

All together, an entrepreneurial society and economy require tax policies that encourage the formation of capital.

Surely one "secret" of the Japanese is their officially encouraged "tax evasion" on capital formation. Legally a Japanese adult is allowed *one* medium-sized savings account the interest on which is tax-exempt. Actually Japan has five times as many such accounts as there are people in the country, children and minors included. This is, of course, a "scandal" against which newspapers and politicians rail regularly. But the Japanese are very careful not to *do* anything to "stop the abuse." As a result they have the world's highest rate of capital formation. This may be considered too circuitous a way to escape the dilemma of modern society: the conflict between the need for capital formation at a high rate and the popular condemnation of interest and dividends as "unearned income" and "capitalist," if not as sinful and wicked. But one way or another any country that wants to remain competitive in an entrepreneurial era will have to develop tax policies which do what the Japanese do by means of semi-official hypocrisy: encourage capital formation.

Just as important as tax and fiscal policies that encourage entrepreneurship—or at least do not penalize it—is protection of the new venture against the growing burden of governmental regulations, restrictions, reports, and paperwork. My own prescription, though I have no illusion of its ever being accepted, would be to allow the new venture, whether an independent enterprise or part of an existing one, to charge the government for the costs of regulations, reports, and paperwork that exceed a certain proportion (say 5 percent) of the new venture's gross revenues. This would be particularly helpful to new ventures in the public-service sector—for example, a freestanding clinic for ambulatory surgery. In developed countries public-service institutions are even more heavily burdened by governmental red tape, and even more loaded down with doing chores for the government than are businesses. And they are even less able, as a rule, to shoulder the burden whether in money or in people.

Such a policy, by the way, would be the best—perhaps the only—remedy for that dangerous and insidious disease of developed countries: the steady growth in the invisible cost of government. It is a real cost in money and, even more, in capable people, their time, and their efforts. The cost is invisible, however, since it does not show in governmental budgets but is hidden in the accounts of the physician whose nurse spends half her time filling out governmental forms and reports, in the budget of the university where sixteen high-level administrators work on "compliance" with governmental mandates and regulations, or in the profit-and-loss statement of the small business nineteen of whose 275 employees, while being paid by the company, actually work as tax collectors for the government, deducting taxes and Social Security contributions from the pay of their fellow workers, collecting tax-identification numbers of suppliers and customers and reporting them to the government, or, as in Europe, collecting value-added-tax (VAT).

And these invisible governmental overheads are totally unproductive. Does anyone, for instance, believe that tax accountants contribute to national wealth or to productivity, and altogether add to society's well-being, whether material, physical or spiritual? And yet in every developed country government mandates misallocation of a steadily growing portion of our scarcest resource, able, diligent, trained people, to such essentially sterile pursuits.

It may be too much to hope that we can arrest—let alone excise—the cancer of government's invisible costs. But at least we should be able to protect the new entrepreneurial venture against it.

We need to learn to ask in respect to any proposed new governmental policy or measure: Does it further society's ability to innovate? Does it promote social and economic flexibility? Or does it impede and penalize innovation and entrepreneurship? To be sure, impact on society's ability to innovate cannot and should not be the determining, let alone the sole criterion. But it needs to be taken into consideration before a new policy or a new measure is enacted—and today it is not taken into account in any country (except perhaps in Japan) or by any policy maker.

V

THE INDIVIDUAL IN ENTREPRENEURIAL SOCIETY

In an entrepreneurial society individuals face a tremendous challenge, a challenge they need to exploit as an opportunity: the need for continuous learning and relearning.

In traditional society it could be assumed—and was assumed—that learning came to an end with adolescence or, at the latest, with adulthood. What one had not learned by age twenty-one or so, one would never learn. But also what one had learned by age twenty-one or so one would apply, unchanged, the rest of one's life. On these assumptions traditional apprenticeship was based, traditional crafts, traditional professions, but also the traditional systems of education and the schools. Crafts, professions, systems of education, and schools are still, by and large, based on these assumptions. There were, of course, always exceptions, some groups that practiced continuous learning and relearning: the great artists and the great scholars, Zen monks, mystics, the Jesuits. But these exceptions were so few that they could safely be ignored.

In an entrepreneurial society, however, these "exceptions" become the exemplars. The correct assumption in an entrepreneurial society is that individuals will have to learn new things well after they have become adults—and maybe more than once. The correct assumption is that what individuals have learned by age twenty-one will begin to become obsolete five to ten years later and will have to be replaced— or at least refurbished—by new learning, new skills, new knowledge.

One implication of this is that individuals will increasingly have to take responsibility for their own continuous learning and relearning, for their own self-development and for their own careers. They can no longer assume that what they have learned as children and youngsters will be the "foundation" for the rest of their lives. It will be the "launching pad"—the place to take off from rather than the place to build on and to rest on. They can no longer assume that they "enter upon a career" which then proceeds along a pre-determined, well-mapped and well-lighted "career path" to a known destination—what the American military calls "progressing in grade." The assumption from now on has to be that individuals on their own will have to find, determine, and develop a number of "careers" during their working lives.

And the more highly schooled the individuals, the more entrepreneurial their careers and the more demanding their learning challenges. The carpenter can still assume, perhaps, that the skills he acquired as apprentice and journeyman will serve him forty years later. Physicians, engineers, metallurgists, chemists, accountants, lawyers, teachers, managers had better assume that the skills, knowledges, and tools they will have to master and apply fifteen years hence are going to be different and new. Indeed they better assume that fifteen years hence they will be doing new and quite different things, will have new and different goals and, indeed, in many cases, different "careers." And only they themselves can take responsibility for the necessary learning and relearning, and for directing themselves. Tradition, convention, and "corporate policy" will be a hindrance rather than a help.

This also means that an entrepreneurial society challenges habits and assumptions of schooling and learning. The educational systems the world over are in the main extensions of what Europe developed in the seventeenth-century. There have been substantial additions and modifications. But the basic architectural plan on which our schools and universities are built goes back three hundred years and more. Now new, in some cases radically new, thinking and new, in some cases radically new, approaches are required, and on all levels. Using computers in preschool may turn out to be a passing fad. But four-year-olds exposed to television expect, demand, and respond to very different pedagogy than four-year-olds did fifty years ago. Young people headed for a "profession"—that is, four-fifths of today's college students—do need a "liberal education." But that clearly means something quite different from the nineteenth-century version of the seventeenth-century curriculum that passed for a "liberal education" in the English-speaking world or for *"Allgemeine Bildung"* in Germany. If this challenge is not faced up to, we risk losing the fundamental concept of a "liberal education" altogether and will descend into the purely vocational, purely specialized, which would endanger the educational foundation of the community and, in the end, community itself. But also educators will have to accept that schooling is not for the young only and that the greatest challenge—but also the greatest opportunity—for the school is the continuing relearning of already highly schooled adults.

So far we have no educational theory for these tasks. So far we have no one who does what, in the seventeenth century, the great Czech educational reformer Johann Comenius did or what the Jesuit educators did when they developed what to this day is the "modern" school and the "modern" university. But in the United States, at least, practice is far ahead of theory. To me the most positive development in the last twenty years, and the most encouraging one, is the ferment of educational experimentation in the United States—a happy by-product of the absence of a "Ministry of Education"—in respect to the continuing learning and relearning of adults, and especially of highly schooled professionals. Without a "master plan," without "educational philosophy," and, indeed, without much support from the educational establishment, the continuing education and professional development of already highly educated and highly achieving adults has become the true "growth industry" in the United States in the last twenty years.

The emergence of the entrepreneurial society may be a major turning point in history.

A hundred years ago, the worldwide panic of 1873 terminated the Century of Laissez-Faire that had begun with the publication of Adam

Smith's *The Wealth of Nations* in 1776. In the Panic of 1873 the modern welfare state was born. A hundred years later it had run its course, almost everyone now knows. It may survive despite the demographic challenges of an aging population and a shrinking birthrate. But it will survive only if the entrepreneurial economy succeeds in greatly raising productivities. We may even still make a few minor additions to the welfare edifice, put on a room here or a new benefit there. But the welfare state is past rather than future—as even the old liberals now know.

Will its successor be the Entrepreneurial Society?

Suggested Readings

Most of the literature on entrepreneurship is anecdotal and of the "Look, Ma, no hands" variety. The best of that genre may be the book by George Gilder: *The Spirit of Enterprise* (New York: Simon & Schuster, 1984). It consists mainly of stories of individuals who have founded new businesses; there is little discussion of what one can learn from their example. The book limits itself to new small businesses and omits discussion of entrepreneurship in both the existing business and the public-service institution. But at least Gilder does not make the mistake of confining entrepreneurship to high tech.

Far more useful to the entrepreneur—and to those who want to understand entrepreneurship—are the studies by Karl H. Vesper of the University of Washington in Seattle, Washington, especially his *New Venture Strategy* (Englewood Cliffs, N.J.: Prentice-Hall, 1980), and his annual publication, *Frontiers of Entrepreneurship Research* (Babson Park, Mass.: Babson College). Vesper, too, confines himself to the new and especially to the small business. But within these limits, his stimulating works are full of insights and practical wisdom.

The Center for Entrepreneurial Management (83 Spring Street, New York, N.Y. 10012), founded and directed by Joseph R. Mancuso, focuses entirely on "How to Do It" in the small business, as does Mancuso's well-known text *How to Start, Finance and Manage Your Own Small Business* (Englewood Cliffs, N.J.: Prentice-Hall, 1978).

Entrepreneurial management in the existing and especially in the large business is the subject of two very different books that complement each other. Andrew S. Grove, one of the founders and now the president of Intel Corporation, discusses the policies and practices needed to maintain entrepreneurship in the business that has grown fast and to large size in his *High-Output Management* (New York: Random House, 1983). Rosabeth M. Canter, an organizational psycholo-

267

gist at Yale University, discusses the attitudes and behavior of corporate leaders in entrepreneurial companies in her book *The Change Masters* (New York: Simon & Schuster, 1983). By far the most penetrating discussion of entrepreneurship in existing businesses is the almost inaccessible article by two members of the consulting firm of McKinsey & Company, Richard E. Cavenaugh and Donald K. Clifford, Jr.: "Lessons from America's Mid-Sized Growth Companies," *McKinsey Quarterly* (Autumn 1983). Publication of a book by the same authors, based on the article and the study on which it reports, is expected in 1985 or 1986.

Of the many books on strategy, the most useful may be Michael Porter's *Competitive Strategies* (New York: Free Press, 1980).

In my own earlier works, entrepreneurship and entrepreneurial management are discussed in *Managing for Results* (New York: Harper & Row, 1964), especially Chapters 1–5, and in *Management: Tasks, Responsibilities, Practices* (New York: Harper & Row, 1973), Chapters 11–14 (The Service Institution) and Chapters 53–61 (Strategies and Structures).

Index

BOOKS BY PETER F. DRUCKER

THE DAILY DRUCKER
366 Days of Insight and Motivation for Getting the Right Things Done

ISBN 0-06-074244-5 (hardcover)
366 daily readings harvested from Drucker's lifetime of work.

THE EFFECTIVE EXECUTIVE
ISBN 0-06-051607-0 (paperback)
Demonstrates the distinctive skill of the executive and offers fresh insights into old and seemingly obvious business situations.

THE ESSENTIAL DRUCKER
In One Volume the Best of Sixty Years of Peter Drucker's Essential Writings on Management

ISBN 0-06-093574-X (paperback)
ISBN 0-06-621087-9 (hardcover)
Covers the basic principles and concerns of management and its problems, challenges, and opportunities.

MANAGEMENT
Tasks, Responsibilities, Practices

ISBN 0-887-30615-2 (paperback)
Equips the manager with the understanding, the thinking, the knowledge, and the skills for today's and also tomorrow's jobs.

MANAGING FOR RESULTS
ISBN 0-887-30614-4 (paperback)
Combines specific economic analysis with a grasp of the entrepreneurial force in business prosperity.

MANAGING IN TURBULENT TIMES
ISBN 0-887-30616-0 (paperback)
Focuses on entering a new economic era with new trends, new markets, new currencies, new principles, new technologies, and new institutions.

MANAGING THE NON-PROFIT ORGANIZATION
Principles and Practices

ISBN 0-887-30601-2 (paperback)
Gives examples and explanations of mission, leadership, resources, marketing, goals, people development, decision making, and much more.

POST-CAPITALIST SOCIETY
ISBN 0-887-30661-6 (paperback)
Provides an incisive analysis of the major world transformation taking place.

MANAGEMENT CHALLENGES FOR THE 21ST CENTURY
ISBN 0-887-30999-2 (paperback)
ISBN 0-887-30998-4 (hardcover)
ISBN 0-694-52212-0 (audio)
How the new paradigms of management have changed and will continue to change our basic assumptions about the practices and principles of management.

THE PRACTICE OF MANAGEMENT
ISBN 0-887-30613-6 (paperback)
ISBN 1-559-94278-9 (audio)
The first book to depict management as a distinct function and to recognize managing as a separate responsibility, this classic work by Peter Drucker is the fundamental and basic book for understanding these ideas.

THE EXECUTIVE IN ACTION
Three Drucker Management Books on What to Do and Why and How to Do It

ISBN 0-887-30828-7 (hardcover)
Three complete Drucker management books in one volume —*Managing for Results, Innovation and Entrepreneurship,* and *The Effective Executive* with a new preface by the author.

All titles are also available in ebooks from PerfectBound

Don't miss the next book by your favorite author.
Sign up for AuthorTracker by visiting *www.AuthorTracker.com*.

Available wherever books are sold, or call 1-800-331-3761 to order.